Lecture Notes in Mathematics

A collection of informal reports and seminars
Edited by A. Dold, Heidelberg and B. Eckmann, Zürich

303

Graph Theory and Applications

Proceedings of the Conference at
Western Michigan University, May 10 - 13, 1972

Sponsored jointly by Western Michigan University
and the U. S. Army Research Office-Durham, under
Grant Number DA-ARO-D-31-124-72-G155

Edited by Y. Alavi, D. R. Lick, and A. T. White
Western Michigan University, Kalamazoo, MI/USA

Springer-Verlag
Berlin · Heidelberg · New York 1972

AMS Subject Classifications (1970): 05 C xx, 94 A 20

ISBN 3-540-06096-0 Springer-Verlag Berlin · Heidelberg · New York
ISBN 0-387-06096-0 Springer-Verlag New York · Heidelberg · Berlin

Offsetdruck: Julius Beltz, Hemsbach/Bergstr.

This volume constitutes the proceedings of the Conference on Graph Theory and Applications, held at Western Michigan University in Kalamazoo, May 10-13, 1972. Participants included undergraduate and graduate students, and research mathematicians from colleges, universities, and industry; in all twenty-three states and four countries were represented. The contributions to this volume* include a wide variety of topics of current research in graph theory and of applications of graph theory.

Unless otherwise specified, all graphs will be finite and undirected, without loops or multiple edges; that is, "Michigan graphs." The basic terminology and notation are reasonably consistent with that of Behzad and Chartrand's An Introduction to the Theory of Graphs (Allyn and Bacon, Boston, 1971) or of Harary's Graph Theory (Addison-Wesley, Reading, Mass., 1969). Exceptions and additional terms and notations are clearly explained in the particular article employing them.

---

* P. Erdös and S. Shelah, H. Levinson and E. Rapaport were unable to present their papers at the Conference, but their contributions are included as part of the published record.

ACKNOWLEDGEMENTS

The editors wish to thank the many people who contributed to the success of the Conference and to the preparation of these proceedings. In addition to the Conference speakers and participants and contributors to this volume, we gratefully acknowledge

The financial support of Western Michigan University and its overall support of the Conference

The financial support of U.S. Army Research Office-Durham

The use of facilities of the Department of Mathematics and its overall support of the Conference

The assistance of the Western Michigan University Office of Research Services

The encouragement and the support of Professor A. Bruce Clarke, Chairman, Department of Mathematics

The counsel, encouragement, and the assistance throughout the entire project of our colleague Professor Gary Chartrand

The assistance and encouragement of our graph theory colleagues Professors S.F. Kapoor and R. E. Pippert

The administrative assistance of Mrs. Patricia Williams

The general and secretarial assistance of Mrs. Judith Charron, Mrs. Darlene Lard, and Mrs. Patricia Williams

The invaluable work of a dedicated team of conference assistants who contributed immeasurably to the smooth running of the Conference:

Daniel C. Chang
William J. Goodwin
Tenho Hindert
Paul E. Himelwright
Mary I. Irvin
Linda M. Lesniak
Robert K. Nelson
Donald W. VanderJagt
James E. Williamson

The excellent work of Paul E. Himelwright for the preparation of the drawings in this volume

And, most definitely, special thanks are due for the outstanding work of Mrs. Darlene Lard, who so patiently and skillfully typed the manuscript.

The editors apologize in advance for any oversights in these acknowledgements or any errors in the manuscript.

Y.A.
D.R.L.
A.T.W.

TABLE OF CONTENTS

VIII

Dedicated

to the memory of

Professor J. W. T. Youngs

TRIPARTITE GRAPHS TO ANALYZE THE
INTERCONNECTION OF NETWORKS

William N. Anderson, Jr.
University of Maryland
College Park, MD 20742,

Richard J. Duffin
Carnegie-Mellon University
Pittsburgh, PA 15213,

and

George E. Trapp
West Virginia University
Morgantown, WV 26505

1. Introduction. In electrical network theory, many properties
of connected networks are determined primarily by the connection and
not the particular components that are connected. In this paper, we
begin by viewing the interconnection of networks as a graph defined on
three sets of vertices. By considering the networks as graphs, we are
able to employ the concepts of adjacency matrices. We obtain results
concerning interconnected graphs that are independent of our electrical network model.

Briefly, a "junctor" is a tripartite graph; two of the sets of
vertices will be joined to the terminals of the networks to be connected; the third set will be the terminals of the new network. It
would surely be impossible to develop a theory of network connections
which allowed a completely arbitrary connection between two networks;
moreover, physical considerations impose certain restrictions on the
type of connection one would wish to consider. Our definition of
junctor is motivated primarily by physical considerations; these turn
out to be sufficiently restrictive that we are able to develop a rea-
sonable mathematical theory.

A general physical consideration is that the junctor itself
should not restrict the possible current flows in the networks; thus
we do not allow unconnected terminals or connections between terminals
in the same set (for these would short the corresponding network).
Another physical restriction we impose is that the junctor be "associative", so that when three networks are to be connected, say in the order
A-B-C, it does not matter whether A and B or B and C are connected first.

In Section 2 we consider various problems involved in connecting three networks; Theorem 1 gives a canonical form for associative junctors. The idea of "compatible" junctors is introduced, and it is shown that compatibility is a generalization of associativity.

In Section 3 we consider current flows in junctors. An n-port network is an electrical network with 2n terminals, divided into pairs — called "ports" — such that the current into one terminal of a pair is necessarily equal to the current out of the other terminal. We classify the junctors that preserve the port current behavior, that is, what junctors guarantee that the new network is an n-port, given that the two components are n-port networks?

We discuss possible extensions of this work in Section 4.

Another paper containing the relationship of this work to electrical networks, and also includes the complete proofs of theorems whose proofs are sketched here, has been submitted for publication; see [2].

2. Junctors. A junctor is a graph whose set of vertices is divided into three equal classes, $A = \{a_i\}$, $B = \{b_i\}$, and $C = \{c_i\}$ such that

i. Each vertex is connected to some other vertex.

ii. No vertex is connected to another vertex of its own class.

iii. No vertex is connected to two distinct vertices of the same class.

A junctor is thus specified by the three incidence matrices K, L, and M, where

$$k_{ij} = \begin{cases} 1 & \text{if vertex } a_j \text{ is connected to vertex } c_i, \\ 0 & \text{otherwise,} \end{cases}$$

$$\ell_{ij} = \begin{cases} 1 & \text{if vertex } b_j \text{ is connected to vertex } c_i, \\ 0 & \text{otherwise,} \end{cases}$$

$$m_{ij} = \begin{cases} 1 & \text{if vertex } a_j \text{ is connected to vertex } b_i, \\ 0 & \text{otherwise.} \end{cases}$$

It follows from the definition that K, L, M each have the property that no row or column contains more than one 1. Alternatively,
we may specify the junctor J by the single adjacency matrix.

$$J = \begin{bmatrix} 0 & M^* & K^* \\ M & 0 & L^* \\ K & L & 0 \end{bmatrix}.$$

We will sometimes use the following symbolic representative of
the graph and the adjacency matrices:

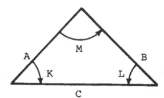

Example 1. A simple example of a junctor is the <u>hybrid junctor</u>
(the name comes from electrical network theory [4], [5], [6], [9])
defined by the graph given below.

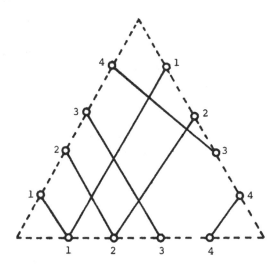

The adjacency matrices for the hybrid junctor are:

$$K = \begin{bmatrix} 1 & 0 & 0 & 0 \\ 0 & 1 & 0 & 0 \\ 0 & 0 & 1 & 0 \\ 0 & 0 & 0 & 0 \end{bmatrix}, \qquad L = \begin{bmatrix} 1 & 0 & 0 & 0 \\ 0 & 1 & 0 & 0 \\ 0 & 0 & 0 & 0 \\ 0 & 0 & 0 & 1 \end{bmatrix},$$

$$M = \begin{bmatrix} 0 & 0 & 0 & 0 \\ 0 & 0 & 0 & 0 \\ 0 & 0 & 0 & 1 \\ 0 & 0 & 0 & 0 \end{bmatrix}.$$

In electrical network literature, the hybrid connection would be represented by the following diagram.

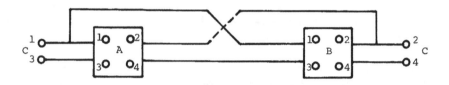

A <u>terminal</u> <u>bank</u> is a device with n terminals. No mathematical properties of this device are assumed here; in the application which motivated this work the terminal bank is an electrical network with n terminals - no doubt other interpretations can be given. Given two terminal banks R, S, and a junctor J, we may form a new terminal bank J(R,S) by identifying the terminals of R with the vertices of class A, the terminals of S with vertices of class B, and calling the vertices of class C the terminals of the new terminal bank.

If we have three terminal banks, R, S, and T, and two copies of the junctor J, we may form the terminal box J(R,J(S,T)) with adjacency matrix

(3) $\qquad$ J(R,J(S,T)) =

	R	S	T	new
R	0	M*K	M*L	K*
S	K*M	0	M*	K*L
T	L*M	M	0	L*2
new	K	LK	L^2	0

.

This connection is shown symbollically in Figure 1.

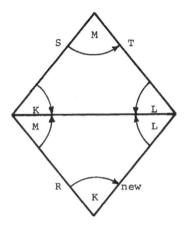

Figure 1. J(R,J(S,T))

Alternatively, we may form the terminal bank J(J(R,S),T) with adjacency matrix

$$J(J(R,S),T) = \begin{array}{c c} & \begin{array}{cccc} R & S & T & new \end{array} \\ \begin{array}{c} R \\ S \\ T \\ new \end{array} & \left[\begin{array}{cccc} 0 & M* & K*M* & K*2 \\ M & 0 & L*M* & L*K* \\ MK & ML & 0 & L* \\ K^2 & KL & L & 0 \end{array}\right] \end{array}$$

(4)

and symbolic representation given in Figure 2. In case J(R,J(S,T)) and J(J(R,S),T) define the same connections between the terminals of R, S, T and the new terminals, we say that the junctor is associative.

Most of the junctors arising in the study of electrical networks are associative, but the non-associative ones are of some interest. An example of an associative junctor is the hybrid junctor of Example 1. We now characterize all associative junctors.

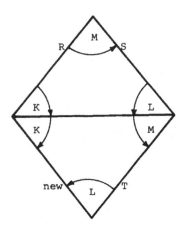

Figure 2. J(J(R,S),T)

THEOREM 1. The junctor J is associative if and only if the
vertices may be renumbered - the same permutation being used in all
three classes - so that the matrices K, L, and M are

$$K = \begin{bmatrix} I & 0 & 0 \\ 0 & I & 0 \\ 0 & 0 & 0 \end{bmatrix}, \qquad L = \begin{bmatrix} I & 0 & 0 \\ 0 & 0 & 0 \\ 0 & 0 & I \end{bmatrix}, \qquad M = \begin{bmatrix} N & 0 & 0 \\ 0 & 0 & P \\ 0 & 0 & 0 \end{bmatrix},$$

where P is a permutation matrix, and N is a diagonal matrix.

Proof. We omit the exact details, but summarize the basic idea.
We want to show that two vertices are connected by a path in
J(R,J(S,T)), if and only if they are connected by a path in
J(J(R,S),T). Since the product of adjacency matrices specifies paths
in a graph and we have the adjacency matrix representations of (3) and
(4), we need only compute the various connections.

The two connections of (3) and (4) may be generalized. Suppose
we have two junctors J_1 and J_2, and the three terminal banks R,
S, and T. Then we may form the terminal banks

$$J_1(R,J_2(S,T)) = \begin{array}{c} \\ R \\ S \\ T \\ \text{new} \end{array} \begin{array}{|c|c|c|c|} \hline R & S & T & \text{new} \\ \hline 0 & M_1{}^*K_2 & M_1{}^*L_2 & K_1{}^* \\ \hline K_2{}^*M_1 & 0 & M_2{}^* & K_2{}^*L_1 \\ \hline L_2{}^*M_1 & M_2 & 0 & L_2{}^*L_1 \\ \hline K_1 & L_1K_2 & L_1L_2 & 0 \\ \hline \end{array} \tag{5}$$

and

$$J_1(J_2(R,S),T) = \begin{array}{c} \\ R \\ S \\ T \\ \text{new} \end{array} \begin{array}{|c|c|c|c|} \hline R & S & T & \text{new} \\ \hline 0 & M_2{}^* & K_2{}^*M_1{}^* & K_2{}^*K_1{}^* \\ \hline M_2 & 0 & L_2{}^*M_1{}^* & L_2{}^*K_1{}^* \\ \hline M_1K_2 & M_1L_2 & 0 & L_1{}^* \\ \hline K_1K_2 & K_1L_2 & L_1 & 0 \\ \hline \end{array} \tag{6}$$

where J_1 is determined by K_1, L_1 and M_1, and J_2 is determined by K_2, L_2 and M_2.

In case $J_1(R,J_2(S,T)) = J_1(J_2(R,S),T)$, then we say J_2 is compatible with J_1. For an example of compatibility, let J_1 and J_2 be defined by Figure 3. It is easy to verify that J_2 is compatible with J_1, but not conversely. Moreover, J_1 is an example of a non-associative junctor, while J_2 is associative.

The following two theorems relate compatible and associative junctors. We omit the proofs. Again the important idea is that the product of adjacency matrices specifies paths.

THEOREM 2. <u>Let</u> J_1 <u>be an associative junctor, and</u> J_2 <u>a junctor such that</u> J_2 <u>is compatible with</u> J_1; <u>then</u> $J_2 = J_1$.

THEOREM 3. <u>Let</u> J_1 <u>and</u> J_2 <u>be junctors such that</u> J_2 <u>is compatible with</u> J_1 <u>and</u> J_1 <u>is compatible with</u> J_2; <u>then</u> $J_1 = J_2$ <u>and this junctor is associative.</u>

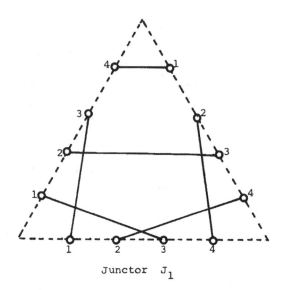

Junctor J_1

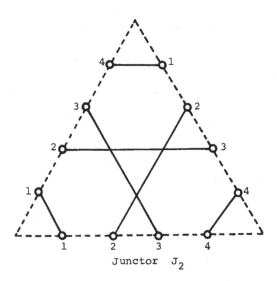

Junctor J_2

Figure 3. Junctor Compatibility

3. _Interconnections of n-port networks_. The electrical networks we wish to interconnect with junctors are known as "n-port networks". Briefly, an _n-port network_ has n pairs (ports) of terminals; the internal structure, which is otherwise irrelevant here, forces the currents to obey the condition that the current into one terminal of a port equals the current out of the other terminal of the same port. This current behavior is sometimes forced by using transformers inside the networks. In this section we discuss network interconnections which avoid the use of such non-graphical devices.

A _iso-flow junctor_ is an associative junctor which forms an n-port network from the interconnection of two n-port networks. That is, the junctor has 6n vertices, and if the networks connected at A and B satisfy the current conditions for n-ports, then so will the interconnected network. Kirchhoff's current law easily leads to an algebraic condition on the adjacency matrices which characterize this property.

The hybrid junctor seen above is an example of a 2-port iso-flow junctor; in fact, the hybrid junctor is formed from the 1-port _parallel_ junctor (at port 1: terminals 1 and 3), and the 1-port _series junctor_ (at port 2). Other examples of 2-port iso-flow junctors are the _cascade junctor_ and the _crossed-cascade junctor_, see Figure 4.

A _hybrid-cascade_ n-port junctor is formed by connecting n_1 ports in parallel, n_2 ports in series, $2n_3$ ports in cascade and $2n_4$ ports in crossed-cascade $(n_1 + n_2 + 2n_3 + 2n_4 = n)$, with no other connection between the various ports. It may be easily seen that a hybrid-cascade junctor is a iso-flow junctor; the following converse holds.

THEOREM 4. _The most general iso-flow junctor is a hybrid-cascade junctor_.

The proof follows from considering the algebraic conditions which characterize resistive junctors; no ideas from electrical network theory, other than the initial use of Kirchhoff's laws, are needed.

The electrical aspects of the hybrid-cascade connections have been studied by us elsewhere [1] - [5], [9]; in view of the above theorem, there are no more purely graphical connections to be considered. However, there exists a broad class of connections using transformers and other non-graphical devices; the study of these has barely been started.

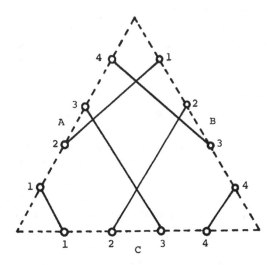

Cascade Junctor

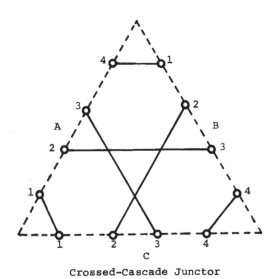

Crossed-Cascade Junctor

Figure 4. Cascade Junctors

4. Generalizations and related topics. For an associative junctor we have seen in Theorem 1 that the matrix N may be any diagonal matrix of 0's and 1's. If the junctor were to represent an electrical network, then the situation would be - a_i connected to c_i and c_i connected to b_i; then a_i and b_i would be electrically connected but a direct wire might or might not connect them. It is easier mathematically to analyze the case where $N = 0$. Another way to state this condition is that the junctor has no embedded triangles.

The physical realization of a junctor is important in electrical network theory. For this reason one would like a characterization of those junctors which are planar when drawn as in the second figure, with possibly a consistent permutation of the vertex numbers.

The two most common interconnections of networks are the series and parallel connections, [1], [5]. Duffin [3] and Riordan and Shannon [8], have characterized series and parallel networks (graphs) as those that do not contain the complete four graph. By joining many terminal banks together with different junctors, one obtains a mixed junctor graph. A characterization of series-parallel mixed junctor graphs would be valuable.

The ideas of compatibility could also be extended to mixed junctors. A characterization of mixed junctors which are obtained by using only compatible (associative) junctors would be useful.

An extension of the concept of a junctors to matroids [7] should also lead to an interesting theory.

References

1. W. N. Anderson, Jr. and R. J. Duffin, Series and parallel addition of matrices, J. Math. Anal. Appl. 26 (1969), 576-594.

2. W. N. Anderson, Jr., R. J. Duffin, and G. Trapp, Incidence matrix concepts for the analysis of the interconnections of networks, TR-71-27, University of Maryland, July 9, 1971.

3. R. J. Duffin, Topology of series-parallel networks, J. Math. Anal. Appl. 10 (1965), 303-318.

4. R. J. Duffin, D. Hazony, and N. Morrison, Network synthesis through hybrid matrices, SIAM J. Appl. Math. 14 (1966), 390-413.

5. R. J. Duffin and G. Trapp, Hybrid addition of matrices - a network theory concept, J. Applicable Analysis, to appear.

6. L. Huelsman, Circuits, Matrices, and Linear Vector Spaces, McGraw-Hill, New York, 1963.

7. G. Minty, On the axiomatic foundations of the theories of directed linear graphs, electrical networks, and network programming, <u>J. Math</u>. <u>Mech</u>. 15 (1966), 485-520.

8. J. Riordan and C. E. Shannon, The number of two-terminal series-parallel networks, <u>J. Math. Phys</u>. 21 (1942), 83-93.

9. G. Trapp, <u>Operator Algebra Related to Network Theory</u>, Ph.D. Dissertation, Carnegie-Mellon University, Pittsburgh, 1970.

MINIMAL REGULAR MAJOR MAPS WITH PROPER 4-RINGS

Ruth A. Bari
George Washington University
Washington, DC 20006

1. **Introduction.** Birkhoff and Lewis [2] have proposed a strong
form of the 4-color conjecture in terms of chromatic polynomials.
They have proved in [2] that in a cubic map of simply-connected regions on a sphere, a proper 2-ring, a proper 3-ring, and a 4-sided
region surrounded by a proper 4-ring are absolutely reducible configurations, where by an _absolutely_ _reducible_ _configuration_ is meant
one whose presence in a map assures us that the Birkhoff-Lewis conjecture holds for the given map if it holds for all maps with fewer
regions than the given map.

A map M_n with n regions on the sphere is said to be _regular_
if it is a cubic map of simply-connected regions without proper 2-
rings or proper 3-rings. A region with 5 or more sides is called a
major _region_, while one with less than 5 sides is called a _minor_
region. The map M_n is said to be a _major_ _map_ if all the regions of
of M_n are major regions. A major map which is regular is called a
regular _major_ _map_. Since every regular map which contains none of the
above mentioned reducible configurations is a regular major map, a
proof that these maps are absolutely reducible would thus confirm the
4-color conjecture. Hence it is of some interest to examine the
structure of regular major maps.

A regular major map which contains no proper 4-rings is called a
4-regular _major_ _map_. In this paper it is proved that every regular
major map with fewer than 20 regions is a 4-regular major map, and
that there are exactly two regular major maps with 20 regions which
contain a proper 4-ring.

2. **Preliminaries.** An _m-ring_ C of a map M_n consists of a
closed curve c without double points which passes successively
through m regions $R_1, R_2, \ldots, R_m$ without passing through any
vertices, and which intersects exactly once each side through which it
passes. The m regions $R_1, R_2, \ldots, R_m$ are said to form the ring
C, and c is called the _defining_ _curve_ of C. The region R_i is

assumed to be contiguous to the region R_{i+1} (i taken modulo m).
In addition, the m-ring C is called a proper m-ring if (i) the regions R_i are distinct, (ii) there are no contacts between noncon-
secutive pairs of regions, (iii) the defining curve c of C intersects exactly m sides, and (iv) no two consecutive regions meet
across a side not intersected by c. The inside and outside of the
ring refer strictly to the parts of the map inside and outside of C.
The ring interior map C^I is the map we get from the map M_n by
replacing all regions of M_n outside of C by a single region. The
ring exterior map C^E is the map we get from the map M_n by replacing all regions of M_n inside of C by a single region. The
strict interior map C^{Int} is the map of all regions of C^I except
the regions of C and the exterior region. The strict exterior map
C^{Ext} is the map consisting of all regions of C^E except the regions
of C and the interior region.

If R is a minor region in a submap of M_n, the deficiency
of R is the number of vertices which must be added to R to make it
a major region. The deficiency of a submap of M_n is the sum of the
deficiencies of its regions.

3. 4-Regularity of regular major maps with fewer than 20
regions.

THEOREM 1. There is a regular major map of 20 regions which
has a proper 4-ring.

Proof. See Figure 1, where the proper 4-ring is determined by
the non-pentagonal regions.

NOTE. In the following map, the number 5 designates a pentagon,
6 a hexagon, etc. The symbol $(n; n_1, n_2, n_3, ...)$ designates that
the corresponding map M_n has n regions of which n_1 are quadrilaterals, n_2 are pentagons, n_3 are hexagons, etc.

THEOREM 2. No regular major map with fewer than 20 regions
has a proper 4-ring.

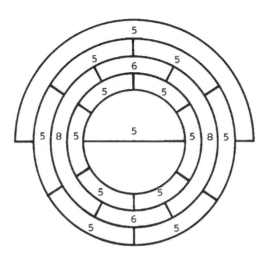

Figure 1.   (20; 0, 16, 2, 0, 2)

Proof.   Let M_n be a regular major map which has a proper 4-ring, and suppose that no regular major map with less than n regions has a proper 4-ring. Then by Theorem 1, n ≤ 20.

Let R_1, R_2, R_3, and R_4 denote the regions which form the 4-ring C, and let C^I be the minimal ring interior map derived from this 4-ring. Then we show that each region of the 4-ring C is a major region of C^I.

We suppose, to the contrary, that a region of C, say R_1, is a quadrilateral of C^J From the definition of a proper ring, it is obvious that every region of C has at least 4 sides. Let c be the defining curve of the 4-ring C, and let v_i be the vertex common to R_i and R_{i+1}, i is an integer modulo 4. Since v_1 is a triple vertex, R_1 and R_2 meet a third region, say S, at v_1. Let ℓ_1 be the boundary between R_1 and S, ℓ_2 be the boundary between R_2 and S which meets v_1. Since R_1 is a quadrilateral, there is no vertex on ℓ_1 between v_1 and v_4, so that $\ell_1 = v_1v_4$ (see Figure 2). Therefore R_1 and R_4 meet S at v_4. Let ℓ_4 be the boundary between R_4 and S. Then there is a defining curve c_1

inside c passing from R_4 to S across ℓ_4, from S to R_2 across ℓ_2, then from R_2 to R_3 and back to R_4. This is possible because R_2, R_3, and R_4 are part of the proper 4-ring C.

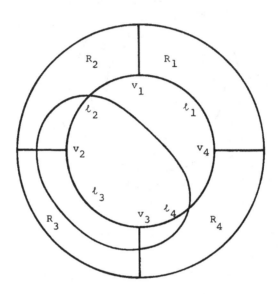

Figure 2.

Furthermore, S cannot be the only region inside of C, because this implies that S is a quadrilateral, and contradicts the assumption that M_n is a major map. Hence, there is a region $S' \neq S$ inside C which meets, say, R_2 at v_2.

It is proved in [1] that every regular major map which has a non-proper 4-ring with a non-vacuous interior and a non-vacuous exterior has at least 29 regions. Since M_n has at most 20 regions, it follows that M_n has no non-proper 4-ring C_1 which has at least one region inside the closed curve c_1 and one region outside c_1. Hence C_1 is a proper 4-ring. But then C_1^I has fewer regions than C^I, contradicting the assumption that C^I is a minimal map of this kind. Therefore R_1 cannot be a quadralateral, and every region of C is a major region.

pentagons. Since c^I has exactly 10 pentagons, two pentagons fail to meet C.

Let S_1 and S_2 be two pentagons which fail to meet C. We now prove that S_1 and S_2 have a common side, and S_2 is in the ring of regions around S_1. This follows because S_1 is surrounded by a 5-ring C_1, so that S_1 and the regions of C_1 add up to 6 regions. If S_2 is not a region of C_1 then S_2 together with the above 6 regions gives us 7 regions in c^{Int}. But it was shown above that if P_n has fewer than 20 regions, c^{Int} has at most 7 regions, so there are no other regions in c^{Int}. The only region inside C_1 is S_1, and S_2 is the only region of c^{Int} outside C_1. If S_2 does not meet C_1 there must be a region of c^{Int} between S_2 and C. This is impossible, because S_2 is the only region of c^{Int} outside C. Hence we have a contradiction, and S_2 is a region of C and meets S_1.

Let C_2 be the ring of regions around the two pentagons which fail to meet C. Then c_2^{Int} is the strict interior map consisting of two regions. Each region of c_2^{Int} has a deficiency of 3. Therefore C_2 has at least 6 regions to delete the deficiency of c_2^{Int}. Because two vertices of c_2^{Int} meet C_2, two regions of C_2 are major regions of c_2^I. Hence the deficiency of c^{Int} is 4, and the four interior vertices of C delete the deficiency of c^{Int}. But then the minimal map c^I, where C is a proper 4-ring, is the map (13; 1, 10, 2). This is precisely the map derived from the 4-ring C in (20; 0, 16, 2, 0, 2).

But then the smallest regular major map which has a proper 4-ring has 20 regions, 13 regions in c^I, 13 regions in c^E, less the 4 regions of C and the two quadrilaterals, one outside C in c^I, the other inside C in c^E.

COROLLARY. Every regular major map with fewer than 20 regions is 4-regular.

THEOREM 3. There are exactly 2 regular major maps M_{20} having a proper 4-ring.

But if c^I is a minimal ring interior map determined by a 4-ring C in a regular major map M_n, c^E may be chosen symmetric to c^I, since it is not necessary for c^E to be different from c^I to make the regions of C become major regions in M_n.

As a consequence of Euler's polyhedral formula, it can be shown that every map on a sphere must satisfy the formula $4f_2 + 3f_3 + 2f_4 + f_5 = 12 + f_7 + 2f_8 + \ldots$, where f_m is the number of m-gons in the map. Since c^I has only one 4-gon, the region outside C, and no regions with fewer than 4 sides, c^I must satisfy $2 + f_5 = 12 + f_7 + 2f_8 + \ldots$. Hence $2 + f_5 \geq 12$, $f_5 \geq 10$, so that there are at least 10 pentagons and one quadrilateral in c^I. If M_n has 20 regions, with 4 regions in C, and c^I symmetric to c^E, there must be 8 regions inside C and 8 regions outside C, so that c^I has $8 + 4 + 1 = 13$ regions. Thus if M_n has fewer than 20 regions, c^I has at most 12 regions, seven in c^{Int}, four in C, and one quadrilateral outside C.

If c^I has a heptagon, $f_4 = f_7 = 1$, so that $2 + f_5 = 12 + 1 = 13$, and $f_5 = 11$. Then c^I has at least 13 regions, a quadrilateral, a heptagon, and 11 pentagons. If c^E and c^I are symmetric in M_n, then M_n has $13 + 13 - 4 - 2 = 20$ regions, since the 4-ring is counted twice instead of once, and the region outside C in c^I and that inside C in c^E does not appear in M_n. Because the inclusion of more regions with seven or more sides would make M_n even larger, we may conclude that if M_n has fewer than 20 regions c^I has no region with more than six sides.

Thus, if M_n has fewer than 20 regions, c^I has exactly 10 pentagons and at most one hexagon (since if c^I has two or more hexagons, c^I has 13 or more regions, so that M_n would have 20 or more regions). In this case, as we prove in the immediate sequel, there are at least two pentagons which do not meet the 4-ring C.

If every region of C is a pentagon in c^I, 4 vertices of c^{Int} meet C in c^I. But then at most four regions of c^{Int} meet C, as proved in [1]. Since there are exactly four pentagons in C and at most four pentagons which meet C in c^I, and since c^I has exactly ten pentagons, at least two pentagons of c^{Int} fail to meet C.

If one region of C is a hexagon and the others are pentagons five vertices of c^{Int} meet C in c^I. Hence at most five regions of c^{Int} meet C in c^I. There are exactly 3 pentagons in C and at most 5 pentagons in c^{Int} which meet C, a total of eight

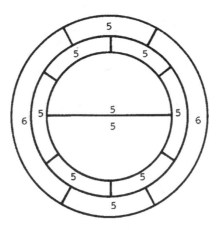

Figure 3.   (13; 1, 10, 2)

Proof. From the discussion in Theorem 2, it follows that
(13; 1, 10, 2) of that theorem is the only map c^I with 13 regions,
and, of course, the same map may be thought of as c^E.

We get maps M_{20} of the desired type only by considering the
union of c^I with c^E, with the 4-ring C coinciding in these maps.
The only way different maps M_{20} can be obtained in this manner is to
rotate c^I relative to c^E or vice versa. Thus, the only way to get
different maps M_{20} is either by allowing the pentagons and hexagons
of C to coincide in this union, or by allowing the pentagons of C
in c^I to coincide with the hexagons of C in c^E. In the first
case, we obtain the map (20; 0, 16, 2, 0, 2) of Theorem 1. In the
second case, we get the map of Figure 4.

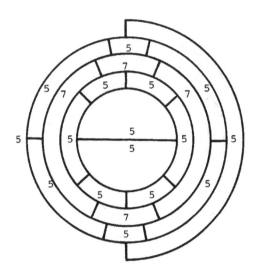

Figure 4.   (20; 0, 16, 0, 4)

References

1. R. A. Bari, *Absolute Reducibility of Maps of at Most 19 Regions*, Ph.D. Thesis, The Johns Hopkins University, 1966 .

2. G. D. Birkhoff and D. C. Lewis, Chromatic polynomials, *Trans. Amer. Math. Soc.* 60 (1946), 355-451.

TOTAL GRAPHS

Mehdi Behzad
Arya-Mehr University
Tehran, Iran

1. **Introduction**. Let G be a graph. Besides the (vertex) chromatic number $\chi(G)$ and the edge chromatic number $\chi_1(G)$, another positive integer $\chi_2(G)$, called the total chromatic number of G, is assigned to G. In analogy with the formula $\chi_1(G) = \chi(L(G))$ the total graph $T(G)$ of G is defined in such a way that $\chi_2(G) = \chi(T(G))$. The structure of regular total graphs is given in [3], and a characterization of nonregular total graphs in [1]. The algorithm provided for determining nonregular total graphs is applicable to regular total graphs as well, with the exceptions of $T(C)$ and $T(K)$, where C is a cycle and K is a nontrivial complete graph. (Definitions not given here may be found in [2], and [5].) In this note we elaborate on the techniques involved in this algorithm.

2. **Preliminaries**. The vertex set of a graph G is denoted by $V(G)$, and its edge set by $E(G)$. The closed neighborhood $\overline{N}_G(v)$ of a vertex v of G is the set consisting of v together with all elements of $V(G)$ adjacent with v.

The total graph $T(G)$ of a graph G has $V(G) \cup E(G)$ as its vertex set and two vertices of $T(G)$ are adjacent if and only if they are adjacent or incident in G. It is easily seen that G and the line-graph $L(G)$ of G are disjoint induced subgraphs of $T(G)$.

In [4] it is shown that two graphs G and H are isomorphic if and only if $T(G)$ and $T(H)$ are isomorphic. There, it was also shown that, apart from isomorphism, G is the only subgraph of $H = T(G)$ whose total graph is H, provided G is a connected graph which is neither a cycle C nor a nontrivial complete graph K. This subgraph, called the special subgraph of H, is denoted by G_s. The subgraph G_s is an induced subgraph every vertex of which is called a special vertex of H, while every other element of $V(H)$ is a nonspecial vertex of H. It is observed that the subgraph induced by the set of nonspecial vertices of H is $L(G_s)$.

If G has m components, then T(G) consists of m components each of which is the total graph of a component of G. The converse is also true. Thus we only consider connected graphs in what follows.

3. <u>Characterization of</u> T(C). Let G be a connected regular graph of degree 4. The only such graph with order 6 or less which is total is $T(K_3)$. This graph is the total graph of every one of its triangles. Thus, we assume that the order of G is at least seven. Then G is the total graph of a cycle if and only if G has even order $2p, V(G)$ is the disjoint union of 2 sets each inducing a cycle C of order p, and $G = T(C)$.

4. <u>Characterization of</u> T(K). Disposing of the trivial cases we assume that the order of G is at least 10. Let $v \in V(G)$, and let H be a complete subgraph of $\langle N_G(v) \rangle$ with maximum order. Then $G = T(K)$ if and only if $G = T(H)$.

5. <u>Characterization of other total graphs</u>. Let $H \neq T(C), T(K)$ be a connected graph and $u \in V(H)$. Then H is a total graph if and only if $H = T(G_{v;u_1v_1})$ for some $v \in \bar{N}_H(u)$ and some edge u_1v_1, where u_1 and v_1 are two even vertices of H which are both adjacent with v, and $G_{v;u_1v_1}$ is constructed below.

If H is a total graph, then some $v \in \bar{N}_H(u)$ is a nonspecial vertex of H and v is adjacent with 2 even adjacent special vertices u_1 and v_1 of H, and the following method enables us to determine the special subgraph $G_{v;u_1v_1}$ of H.

Let $\{i\}$, $i = 0, 1, 2, \ldots, n$, denote the class of all vertices of H whose distance from v is i. Then $n \geq 2$. Next, we separate the sets of special vertices S_i and nonspecial vertices N_i of H which are contained in the class $\{i\}$. Clearly $S_0 = \phi$, $N_0 = \{v\}$, $S_1 = \{u_1, v_1\}$, and $N_1 = \{1\} - S_1 \neq \phi$.

Let $w_1 \in N_1$. Then w_1 is adjacent with one of vertices u_1 and v_1, say u_1, and a special vertex of H, say u_2, contained in $\{2\}$. Necessarily u_2 is adjacent with u_1. Every element of $\{2\}$ which is adjacent with u_1 is a special vertex and, among these vertices only one is adjacent with both u_1 and w_1. Thus u_2 can be uniquely determined. We repeat this argument for all elements of

N_1 to obtain S_2. Then $N_2 = \{2\} - S_2$.

To separate N_3 and S_3, let $w_2 \in N_2$. If w_2 is adjacent with 2 elements of S_2, then every element of $\{3\}$ adjacent with w_2 is a nonspecial vertex of $\{3\}$; otherwise w_2 is adjacent with a special vertex, say u_2, of $\{2\}$ and a special vertex, say u_3, of $\{3\}$. We propose to determine u_3. All vertices of $\{3\}$ adjacent with u_2 are special vertices of $\{3\}$, and among these only one is adjacent with both w_2 and u_2. This vertex is u_3. We repeat this procedure to find S_3. Then $N_3 = \{3\} - S_3$.

Using induction and the above method we determine $S_0, S_1, \ldots,$ and S_n. Then $G_{v;u_1v_1} = \langle S_1 \cup S_2 \cup \ldots \cup S_n \rangle$.

References

1. M. Behzad, A characterization of total graphs, Proc. Amer. Math. Soc. 26 (1970), 383-389.

2. _____, and G. Chartrand, Introduction to the Theory of Graphs, Allyn and Bacon, Boston, 1971.

3. _____, and H. Radjavi, Structure of regular total graphs, J. London Math. Soc. 44 (1969), 433-436.

4. _____, The total group of a graph, Proc. Amer. Math. Soc. 19 (1968), 158-163.

5. F. Harary, Graph Theory, Addison-Wesley, Reading, Mass., 1969.

A CENSUS OF BALL AND DISK DISSECTIONS

L. W. Beineke and R. E. Pippert
Purdue University at Fort Wayne
Fort Wayne, IN 46805

1. Introduction. The objects to be considered here are essentially triangulations of polygons and some 3-dimensional analogues.
It is relatively rare in graph theory that explicit formulas are found for the number of unlabeled objects of a given type, but for disk and ball dissections, such formulas have been found. It is our purpose to present these results here.

As indicated in the interesting historical survey of Brown [3], the study of triangulations of polygons (disk dissections) can be traced back to Euler in 1758. Guy [4] compiled and developed several enumeration results in this area, and it is this work that forms the basis of our discussion of disk dissections. Furthermore, the same techniques will be used to obtain our results on ball dissections.

The same outline will be used in treating both cases. After providing definitions and examples, we will give some enumeration results for rooted (or labeled) dissections. This will be followed by a list of possible automorphisms of a dissection and then (the substantial part) the enumeration of unlabeled dissections by their automorphism groups. The sum of all these will then give the total number of dissections.

2. Disk dissections. A triangulation of a convex polygon, a disk dissection, and a 2-tree embeddable in 2-space can all be considered equivalent objects. We shall use here the term disk dissection and define it inductively by means of inscribed polygons in a closed disk: An inscribed triangle is a disk dissection, and a disk dissection with n + 1 triangles is obtained from one with n triangles by adding a new vertex on the boundary and joining it with edges to its two neighboring vertices. The exterior edges of the polygon and arcs of the circle can be identified in the obvious manner.
Figure 1 indicates the inductive step.

 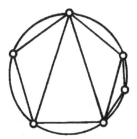

Figure 1. Obtaining a dissection inductively.

The relative distance between consecutive vertices is of course
of no consequence when it comes to distinguishing between dissections,
but it does matter whether rotations and reflections are allowed or
not. Not allowing these transformations gives the same results (up to
a factor) as labeling the vertices, and we call this the "rooted case".
In Figure 2, we show the disk dissections with up to 4 triangles.
The first three can be rooted in 1, 2, and 5 ways, respectively,
and those with four triangles in 6, 6, and 2 ways, respectively.
More generally, we observe that the number of ways of rooting a particular dissection with n triangles is 2(n + 2) divided by the order
of the automorphism group of the dissection.

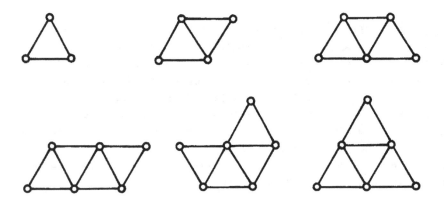

Figure 2. Disk dissections with up to 4 triangles.

There is a useful recursion formula, given in Theorem 1, for the number t(n) of rooted disk dissections with n triangles. It can be used to find the explicit expression for this number given in Theorem 2. For proofs of these results, see [1] or [4].

THEOREM 1. The number t(n) of rooted disk dissections with n triangles satisfies this recursion relation:

$$t(n) = \sum_{i+j=n-1} t(i)t(j) .$$

THEOREM 2. The number t(n) of rooted disk dissections with n triangles is

$$t(n) = \frac{(2n)!}{n!\,(n+1)!} .$$

We now turn to automorphisms of disk dissections. In addition to the identity automorphism, there are those which leave a triangle fixed (in which case at most two are fixed) and those which interchange two triangles with a common edge. There are two of each type, as indicated in the following list. We use Greek letters to indicate the edges of a triangle, with primes denoting the corresponding edges of a second triangle.

AUTOMORPHISMS OF DISK DISSECTIONS

Name	Example
Identity	$(\alpha)(\beta)(\gamma) = \varepsilon$
Reflection	$(\alpha\beta)(\gamma)$
3-cycle	$(\alpha\beta\gamma)$
Inversion	$(\alpha\alpha')(\beta\beta')$
2-cycle	$(\alpha\beta')(\beta\alpha')$

There are seven possible groups for disk dissections: the identity, three others corresponding to groups of permutations of a triangle, and an additional three which correspond to groups of permutations interchanging two triangles with a common edge. (Several of these are abstractly isomorphic.) In the diagrams accompanying the list of groups, an arrow indicates the addition of a rooted dissection along that edge, with similar arrows meaning isomorphic rooted dissections with the same orientation. In the formulas, t(x) is given by

Theorem 2 when x is an integer and is 0 otherwise. As an example of the derivation of the first six formulas, consider f(n). A disk dissection with this group is obtained by taking two rooted dissections and attaching these along the four indicated edges. Taking all possibilities, one gets the indicated sum. However, this sum includes all those with group (g) and furthermore it counts twice all those with group (f), because of the possibility of interchanging the two types of rooted dissections. The use of Theorem 1 simplifies the expression. This is how the first six formulas are derived. Formula (a) is derived from the earlier observation regarding the number of rooted dissections which correspond to the same unrooted one.

g. The group of all automorphisms of order 2:

$$\{\varepsilon, \ (\alpha\beta)(\alpha'\beta'), \ (\alpha\alpha')(\beta\beta'), \ (\alpha\beta')(\alpha'\beta)\}.$$

$$g(n) = t\left(\frac{n-2}{4}\right).$$

f. The group generated by an inversion: $\{\varepsilon, \ (\alpha\alpha')(\beta\beta')\}$.

$$f(n) = \frac{1}{2}\left(\sum_{2i+2j=n-2} t(i)t(j)-g(n)\right)$$

$$= \frac{1}{2}\left(t(\tfrac{n}{2}) - t(\tfrac{n-2}{4})\right).$$

e. The group generated by a 2-cycle: $\{\varepsilon, \ (\alpha\beta')(\alpha'\beta)\}$.

$$e(n) = \frac{1}{2}\left(\sum_{2i+2j=n-2} t(i)t(j)-g(n)\right)$$

$$= \frac{1}{2}\left(t(\tfrac{n}{2})-t(\tfrac{n-2}{4})\right).$$

d. The group of a triangle: $\{\varepsilon, (\alpha\beta), (\alpha\gamma), (\beta\gamma), (\alpha\beta\gamma), (\alpha\gamma\beta)\}$.

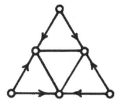

$$d(n) = t\left(\frac{n-4}{6}\right).$$

c. The group generated by a 3-cycle: $\{\varepsilon, (\alpha\beta\gamma), (\alpha\gamma\beta)\}$.

$$c(n) = \frac{1}{2}\left(t\left(\frac{n-1}{3}\right) - d(n)\right)$$

$$= \frac{1}{2}\left(t\left(\frac{n-1}{3}\right) - t\left(\frac{n-4}{6}\right)\right).$$

b. The group generated by a reflection: $\{\varepsilon, (\alpha\beta)\}$.

n odd: $b(n) = t\left(\frac{n-1}{2}\right)$

n even: $b(n) = \frac{1}{2}\left(t\left(\frac{n}{2}\right) - g(n) - 2d(n)\right)$

$$b(n) = t\left(\frac{n-1}{2}\right) + \frac{1}{2}t\left(\frac{n}{2}\right) - \frac{1}{2}t\left(\frac{n-2}{4}\right) - t\left(\frac{n-4}{6}\right).$$

a. The identity group.

$$a(n) = \frac{t(n)}{2n+4} - \frac{b(n)}{2} - \frac{c(n)}{3} - \frac{d(n)}{6} - \frac{e(n)}{2} - \frac{f(n)}{2} - \frac{g(n)}{4}$$

$$= \frac{t(n)}{2n+4} - \frac{1}{2} t(\frac{n-1}{2}) - \frac{3}{4} t(\frac{n}{2}) + \frac{1}{6} t(\frac{n-1}{3}) + \frac{1}{2} t(\frac{n-4}{6}) + \frac{1}{2} t(\frac{n-2}{4}).$$

The sum of these expressions gives the total number of unrooted dissections. The formula is relatively simple.

THEOREM 3. **The number of disk dissections with n triangles is**

$$\frac{1}{2n+4} t(n) + \frac{1}{2} t(\frac{n-1}{2}) + \frac{3}{4} t(\frac{n}{2}) + \frac{1}{3} t(\frac{n-1}{3}) ,$$

where $t(x) = \frac{(2x)!}{x!(x+1)!}$ **if x is an integer and 0 otherwise.**

3. **Ball dissections.** Our definition of **ball dissections** parallels that of disk dissections given earlier, and is made in terms of polyhedra inscribed in a sphere: A regular inscribed tetrahedron is a ball dissection, and given a dissection with n tetrahedra as an inscribed polyhedron, a dissection with n + 1 tetrahedra is obtained by adding a new vertex on the sphere, not lying on the projection from the center of any edge of the polyhedron, and joined to its three neighboring vertices by edges. A decomposition of the closed ball is provided by identification of the sphere and surface of the polyhedron through a projection from the center of the ball. Figure 3 shows schematically the ball dissections with up to four tetrahedra.

Concepts equivalent to ball dissections are 3-trees embeddable in 3-space and dissectible polyhedra (those which can be decomposed by planes meeting only along edges of the polyhedron). We note that whereas a polygon has many dissections, a convex polyhedron has at most one. The familiar regular octahedron is an example of a polyhedron which is not dissectible.

Another approach to ball dissections is through fully decomposable rooted triangulations of the plane: Begin with a fixed triangle and inside triangles successively add vertices joined to each of the three vertices of the triangle. These are called **rooted** ball dissections, because they correspond to ball dissections in which there is a chosen orientation of one edge of a selected exterior triangle. The number of possible ways of rooting a ball dissection is 6 times the number of faces divided by the order of the automorphism group. The numbers of rooted dissections corresponding to the diagrams in Figure 3 are 1, 3, 12, 10, 15, and 30, respectively.

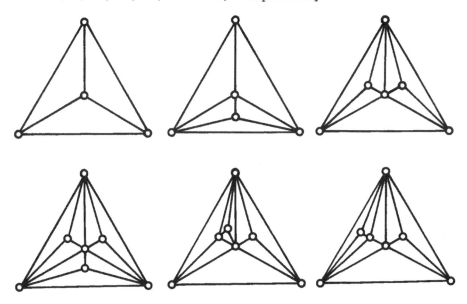

Figure 3. Ball dissections

We denote by $T(n)$ the number of rooted dissections with n tetrahedra, and by $U(n)$ the number of these which are symmetric insofar as they are unchanged when the direction of the root edge is reversed. We take $T(0) \equiv U(0) \equiv 1$. Various recursion formulas for these quantities are derived in [2], and the following result gives some of these.

THEOREM 4. The number $T(n)$ of rooted dissections with n tetrahedra is

$$T(n) = \sum_{i+j+k=n-1} T(i)T(j)T(k).$$

If n is odd, then the number $U(n)$ is

$$U(n) = \sum_{2i+2j=n-1} T(i)T(j) = \frac{3n-1}{n+1}T(\frac{n-1}{2}),$$

while if n is even, then the number U(n) is

$$U(n) = T(\frac{n}{2}).$$

For later convenience, we define another quantity:

$$V(n) = \sum_{i+j=n} U(i)U(j).$$

The next theorem, proved in [1] and [2], gives the values of all these quantities.

THEOREM 5. The numbers T(n), U(n), and V(n) are

$$T(n) = \frac{(3m)!}{n!(2n+1)!} ;$$

$$U(n) = \begin{cases} \dfrac{(3m)!}{m!(2m+1)!} & \text{for} \quad n = 2m, \\[3mm] \dfrac{(3m+1)!}{(m+1)!(2m+1)!} & \text{for} \quad n = 2m + 1; \end{cases}$$

$$V(n) = \begin{cases} \dfrac{(5m+1)(3m)!}{(m+1)!(2m+1)!} & \text{for} \quad n = 2m, \\[3mm] \dfrac{6(3m+2)!}{m!(2m+3)!} & \text{for} \quad n = 2m + 1. \end{cases}$$

Turning now to the automorphisms of ball dissections, we find there are eight. Five of these correspond to permutations of a tetrahedron, and three others interchange two tetrahedra with a common triangle. Here we use Greek letters to indicate faces of tetrahedra: α, β, γ, δ for the four faces of a single tetrahedron, and α, β, γ, and α', β', γ', for the six exterior faces of two tetrahedra with a common triangle, with α and α', β and β', and γ and γ'

having common edges.

AUTOMORPHISMS OF BALL DISSECTIONS

Name	Example
Identity	$\varepsilon = (\alpha)(\beta)(\gamma)(\delta)$
Reflection	$(\alpha\beta)(\gamma)(\delta)$
Digonal rotation	$(\alpha\beta)(\gamma\delta)$
Trigonal rotation	$(\alpha\beta\gamma)(\delta)$
Tetragonal rotation	$(\alpha\beta\gamma\delta)$
Reversal	$(\alpha\alpha')(\beta\beta')(\gamma\gamma')$
Half-turn	$(\alpha\beta')(\beta\alpha')(\gamma\gamma')$
Hexagonal rotation	$(\alpha\beta'\gamma\alpha'\beta\gamma')$

There are seventeen types of automorphism groups of ball dissections (although not all are distinct as abstract groups). These are
listed below, where we also provide a diagram indicating the group,
as well as a formula for the number of ball dissections with that
particular group. Diagrams C and D as well as L - Q represent
the six faces of two tetrahedra which have been cut along certain
edges. Similarly, diagrams E - I represent the four faces of a
single tetrahedron, while B is a single face and J and K give the
twelve exterior faces obtained by adding one tetrahedron to each face
of an initial one. In producing ball dissections of a particular type,
rooted dissections are added to exterior faces in the indicated manner:
For a particular symbol on a face, a symmetric or an arbitrary rooted
dissection is added according as the symbol does or does not have
symmetry. To faces with like symbols, isomorphic rooted dissections
are added in a corresponding manner.

We will indicate here the derivation of only a couple of the formulas because a complete proof would be too lengthy. Derivations of
all formulas will be found in [2].

The case Q is elementary in that the same symmetric rooted dissection is added to six faces, so $Q(n) = U(\frac{n-2}{6})$. The case M is
somewhat more complicated. One can take three rooted dissections on
each of two faces, so one has the sum $\Sigma\, T(i)T(j)T(k)$ with $i + j + k = \frac{1}{2}(n-2)$, which is simply $T(\frac{n}{2})$ by an identity in Theorem 4. This includes several other types of dissections however, as each of types P
and Q is counted once and each of type Θ is counted twice because
of its orientation. This same orientation in M itself means we have
counted everything there twice. The final formula for M(n) results

in, and can of course be replaced by, an expression involving only expressions in T, U, V. As in the disk case, the ball dissections with the identity group are counted by subtracting from the total number of rooted dissections the number of ways of rooting all dissections with a group of higher order. Summing the cases A - Q gives the theorem for the number of unrooted ball dissections.

Q. The group of order 12 generated by a hexagonal rotation and a reflection: $\langle(\alpha\beta'\gamma\alpha'\beta\gamma'),\ (\alpha\beta)(\alpha'\beta')\rangle$.

$$Q(n) = U(\tfrac{n-2}{6}).$$

P. The group of order 4 generated by a reflection and a reversal: $\langle(\alpha\beta)(\alpha'\beta'),\ (\alpha\alpha')(\beta\beta')(\gamma\gamma')\rangle$.

$$P(n) = U(\tfrac{n}{2}) - Q(n).$$

Θ. The group of order 6 generated by a trigonal rotation and a half-turn: $\langle(\alpha\beta\gamma)(\alpha'\beta'\gamma'),\ (\alpha\alpha')(\beta\gamma')(\gamma\beta')\rangle$.

$$\Theta(n) = \tfrac{1}{2}\left(T(\tfrac{n-2}{6}) - Q(n)\right).$$

N. The group of order 6 generated by a hexagonal rotation:

$$\langle (\alpha\beta'\gamma \; \alpha'\beta\gamma') \rangle.$$

$$N(n) = \frac{1}{2}\left(T(\frac{n-2}{6}) - Q(n) \right).$$

M. The group of order 2 generated by a half-turn:

$$\langle (\alpha\beta')(\beta\alpha')(\gamma\gamma') \rangle.$$

$$M(n) = \frac{1}{2}\left(T(\frac{n}{2}) - 2\Theta(n) - P(n) - Q(n) \right).$$

L. The group of order 2 generated by a reversal:

$$\langle (\alpha\alpha')(\beta\beta')(\gamma\gamma') \rangle.$$

$$L(n) = \frac{1}{6}\left(T(\frac{n}{2}) - 2N(n) - 3P(n) - Q(n) \right).$$

K.   The group of order 24 (isomorphic to the symmetric group on four elements) generated by a tetragonal rotation and a trigonal rotation: $\langle (\alpha\beta\gamma\delta), (\alpha\beta\gamma) \rangle$.

$$K(n) = U\left(\frac{n-5}{12}\right).$$

J.   The group of order 12 (isomorphic to the alternating group on four elements) generated by a trigonal rotation and a digonal rotation: $\langle (\alpha\beta\gamma), (\alpha\beta)(\gamma\delta) \rangle$.

$$J(n) = \frac{1}{2}\left(T\left(\frac{n-5}{12}\right) - K(n)\right).$$

I.   The group of order 8 (isomorphic to the dihedral group on four elements) generated by a tetragonal rotation and a reflection: $\langle (\alpha\beta\gamma\delta), (\alpha\beta)(\gamma\delta) \rangle$.

$$I(n) = U\left(\frac{n-1}{4}\right) - K(n).$$

H.   The group of order 4 generated by a tetragonal rotation:

$$\langle (\alpha\beta\gamma\delta) \rangle.$$

$$H(n) = \frac{1}{2}\left(T(\frac{n-1}{4}) - I(n) - K(n)\right).$$

G.   The group of order 4 generated by two digonal rotations:

$$\langle (\alpha\beta)(\gamma\delta), \ (\alpha\gamma)(\beta\delta) \rangle.$$

$$G(n) = \frac{1}{6}\left(T(\frac{n-1}{4}) - 3I(n) - 2J(n) - K(n)\right)$$

F.   The group of order 4 generated by two reflections:

$$\langle (\alpha\beta), \ (\gamma\delta) \rangle.$$

$$F(n) = \frac{1}{2}\left(\sum_{2i+2j=n-1} U(i)U(j) - I(n) - K(n)\right)$$

$$= \frac{1}{2}\left(V(\frac{n-1}{2}) - I(n) - K(n)\right).$$

E. The group of order 2 generated by a digonal rotation:

$$\langle (\alpha\beta)(\gamma\delta) \rangle.$$

For n even, $E(n) = 0.$

For n odd, $E(n) = \frac{1}{4}\Big(\sum_{2i+2j=n-1} T(i)T(j) -$

(those counted in larger groups)$\Big)$

$$= \frac{1}{4}\Big(U(n) - 2F(n) - 6G(n)$$
$$-2H(n) - 3I(n) - 2J(n) - K(n)\Big).$$

D. The group of order 6 (isomorphic to the symmetric group on three elements) generated by a trigonal rotation and a reflection: $\langle (\alpha\beta\gamma), (\alpha\beta) \rangle.$

For $n \equiv 2 \pmod 3$,

$$D(n) = \frac{1}{2}\Big(\sum_{3i+3j=n-2} U(i)U(j) - 2K(n) - Q(n) \Big)$$

$$= \frac{1}{2}\Big(V(\tfrac{n-2}{3}) - 2K(n) - Q(n) \Big).$$

For $n \equiv 1 \pmod 3$,
$$D(n) = U(\tfrac{n-1}{3}).$$
For $n \equiv 0 \pmod 3$,
$$D(n) = 0.$$
In general
$$D(n) = U(\tfrac{n-1}{3}) + \frac{1}{2}\Big(V(\tfrac{n-2}{3}) - 2K(n) - Q(n) \Big).$$

C. The group of order 3 generated by a trigonal rotation:

$$\langle (\alpha\beta\gamma) \rangle.$$

For $n \equiv 2 \,(\text{mod } 3)$,

$$C(n) = \frac{1}{4}\Big(\sum_{3i+3j=n-2} T(i)\,T(j) - 2D(n) - 4J(n)$$

$$-2K(n) - 2N(n) - 2\Theta(n) - Q(n) \Big).$$

For $n \equiv 1 \,(\text{mod } 3)$,

$$C(n) = \frac{1}{2}\Big(T(\tfrac{n-1}{3}) - D(n) \Big).$$

In general,

$$C(n) = \frac{1}{2}\Big(T(\tfrac{n-1}{3}) - D(n) \Big) + \frac{1}{4}\Big(U(\tfrac{2n-1}{3}) - 4J(n) - 2K(n) - 2N(n)$$

$$- 2\Theta(n) - Q(n) \Big).$$

B. The group of order 2 generated by a reflection: $\langle (\alpha\beta) \rangle$.

$$B(n) = \frac{1}{2}\Big(U(n) - 2D(n) - 2F(n) - I(n) \Big)$$

$$- K(n) - P(n) - Q(n) \Big).$$

A. The identity group.

$$A(n) = \frac{T(n)}{12(n+1)} - \sum_{X \neq A} \frac{X(n)}{\sigma(X)},$$

where the sum is over all sixteen other types of automorphism groups
$(X = B, C, \ldots, Q)$, $\sigma(X)$ is the order of that group, and $X(n)$ is
the number of ball dissections with that group.

The sum of these seventeen expressions simplifies rather more than one might expect, and is given in our concluding result:

THEOREM 6. The number $S(n)$ of ball dissections with n tetrahedra is

$$S(n) = \frac{1}{12(n+1)} T(n) + \frac{5}{24} T(\frac{n}{2}) + \frac{1}{3} T(\frac{n-1}{3}) + \frac{1}{4} T(\frac{n-1}{4})$$

$$+ \frac{1}{6} T(\frac{n-2}{6}) + \frac{3}{8} U(n) + \frac{1}{6} U(\frac{2n-1}{3}).$$

where a term $T(m)$ or $U(m)$ is 0 if m is not an integer; and if m is an integer,

$$T(m) = U(2m) = \frac{(3m)!}{m!(2m+1)!},$$

$$U(2m+1) = \frac{(3m+1)!}{(m+1)!(2m+1)!}.$$

References

1. L. W. Beineke and R. E. Pippert, The number of labeled dissections of a k-ball, Math. Ann. 191 (1971), 87-98.

2. _____, Enumerating dissectible polyhedra, submitted for publication.

3. W. G. Brown, Historical note on a recurrent combinatorial problem, Amer. Math. Monthly 72 (1965), 973-977.

4. R. K. Guy, Dissecting a polygon into triangles, Bull. Malayan Math. Soc. 5 (1958), 57-60; Same title, Research Paper No. 9, The University of Calgary, 1967.

NETWORK MODELS FOR MAXIMIZATION OF
HEAT TRANSFER UNDER WEIGHT CONSTRAINTS

S. Bhargava and R. J. Duffin
Carnegie-Mellon University
Pittsburgh, PA 15213

A common problem of heat transfer is the design of machinery so that the structure can dissipate excess heat. For example, cooling fins are used on the cylinders of air cooled engines. Suppose in this example that the fin is not permitted to exceed a given weight. Then an optimum design problem is to find how the thickness of the fin should taper so that the rate of heat dissipation is a maximum.

The cooling fin design problem was solved for circular cylinders by R. J. Duffin in [3] and for convex cylinders by R. J. Duffin and D. K. McLain in [4]. They employed the calculus of variations to recast the question into a max-max problem in [3] and a saddle point problem in [4]. Such variational principles led to the following key lemma -- for an optimum fin the magnitude of the temperature gradient is constant. Using this lemma it is then easy to obtain explicit expressions for the thickness function of the optimum fin.

In this paper an electrical network model for such cooling devices is formulated and studied. Thus consider a lumped network having a finite number of conducting branches. Certain branches, termed set B, are allowed to vary their conductance but the total conductance is limited by the following ℓ_ρ norm type constraint

$$\left(\sum_B g_s^\rho \right)^{1/\rho} \leqslant K. \tag{I}$$

Here g_s is the conductance of branch s and K and ρ are positive constants. Then the design problem is to maximize the joint conductance of the network between two specified input points. Thus it is desired to find Γ, the maximum conductance subject to constraint I.

By a variational argument we establish the following key lemma -- for an optimum network the branch voltages v_s satisfy

(II)
$$|v_s| = \lambda (g_s)^{\frac{p-1}{2}}, \; s \in B.$$

Here λ is a constant and $g_s \neq 0$.

Our network question may be characterized as a maximizing problem of mathematical programming. This suggests that there is a dual minimizing problem. Pursuing this idea leads to the following duality inequality: if $p > 1$,

(III)
$$\|v\|_{2,\alpha} \geq \Gamma^{1/2} \geq 1/\|y\|_{2,\beta}.$$

Here $\| \; \|_{2,\alpha}$ and $\| \; \|_{2,\beta}$ are certain dual norms. The vector v is an arbitrary normalized voltage distribution satisfying Kirchhoff's voltage law. The vector y is an arbitrary normalized current distribution satisfying Kirchhoff's current law. There is no "duality gap"; in other words the duality inequality could be used to give a sharp estimate of Γ. (See [2].)

In a previous paper [1] R. J. Duffin employed the same network model but confined attention to the linear constraint expressed by (I) with $p = 1$. In that paper relations were obtained corresponding to (II) and (III) when p is given the value 1. Presumable there are analogous results for more general nonlinear constraints. However we feel that the ℓ_p constraint is sufficiently important to be singled out for special treatment.

We then return to the cooling fin problem and introduce a L_p constraint analogous to (I). Then reasoning by analogy suggests the form of the key lemma (II) and the duality inequality (III) for the cooling fin. (See forthcoming Networks Journal for details.)

References

1. R. J. Duffin, Optimum heat transfer and network programming, J. Math. Mech. 17 (1968), 759-768.

2. _____, Duality inequalities of mathematics and science, Nonlinear Programming, (edited by J. B. Rosen, O. L. Mangasarian and K. Ritter), Academic Press Inc., New York, 1970, 401-423.

3. _____, A variational problem relating to cooling fins, J. Math. Mech. 8 (1959), 47-56.

4. _____, and D. K. McLain, Optimum shape of a cooling fin on a convex cylinder, J. Math. Mech. 17 (1968), 769-784.

THE "GRAPH THEORY" OF THE GREEK ALPHABET

J. A. Bondy
University of Waterloo
Waterloo, Ontario, Canada

1. Introduction. "When in Michigan do as Michiganders do" is a
wise motto, and I have here drawn inspiration from that well-known
Michigander, Frank Harary. In fact I have plagiarized Professor
Harary for both the title and the basic idea of this paper. The talk
he presented at the Waterloo combinatorics conference in 1968 was
entitled "The Greek Alphabet of Graph Theory" (see [1]); moreover, in
a paper called "Typographs" [2], Professor Harary has already written
about the graph theory of the Roman alphabet.

The idea of taking non-graphical but graphic objects and looking
at them from the point of view of graph theory is not new. Aside from
Harary, Anthony Hill did this with the paintings of Mondrian and discovered that most of these paintings have asymmetric graphs (see [3]).
Let us examine the Greek alphabet from this aspect. (I am told that
Saunders Maclane has discussed the topology of the alphabet in class.)
Take a letter at random, say B. To obtain the graph of this letter
we insert a vertex at each point where two or more lines with differing slopes meet, and at each terminal point of a line. Thus we see
that the graph of the letter B is $K_4 - x$ (see Figure 1).

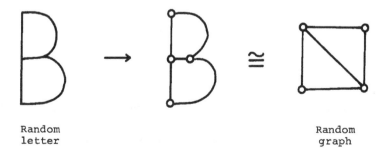

Random
letter

Random
graph

Figure 1.

If we start with the letter θ, we naturally obtain a theta graph, and so on. (By convention, the graph of the letter O is the loop graph.) Clearly, whatever we start with, the resulting graph is always planar. Of what use is it? I shall indicate (rather sketchily) some applications of this concept.

2. **Which graphs are missing?** If one investigates all the letters occurring in alphabets from the early Semitic to the contemporary Roman (Table 1) and those that occur in runic alphabets (Table 2), one finds that certain simple graphs are not represented at all (for example see Figure 2(a)) and that others, which were represented by letters at one time, have since disappeared. An instance of this latter phenomenon is the graph $K_{1,3} + x$ (see Figure 2(b)), which occurs in no alphabet after the early Ionian. Presumably there is some erudite explanation for this, although I shall indicate later (in Section 6) why $\not{y}$ has the characteristics of a relatively 'good' letter.

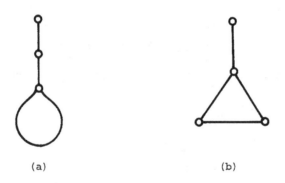

(a)   (b)

Figure 2.

It is interesting, in this regard, to note that all trees on six or fewer vertices do appear as graphs of known letters or symbols. The period, of course yields the one-vertex tree, and the only other trees of order six or less which are not graphs of Roman letters are illustrated in Figure 3.

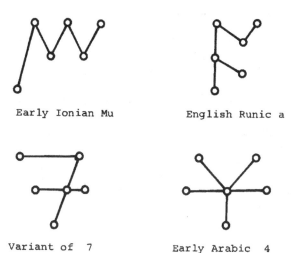

Early Ionian Mu	English Runic a
Variant of 7	Early Arabic 4

Figure 3.

3. **Development of alphabets.** One may follow the development of
the alphabet from its beginning (2000-1500 B.C.) and find out graphi-
cally how individual letters have been transformed to their present
Roman (or Cyrillic) form. The letter A in its various stages is
shown in Figure 4(a); the development of the Arabic numerals 1, 2, 3,
and 4 is illustrated in Figure 4(b).

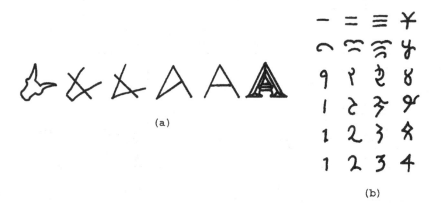

(a)

(b)

Figure 4.

#	early Semitic signs	early Hebrew pen letters	modern Hebrew names	early Ionian signs	classical Greek signs	modern Greek signs	modern Greek names	Etruscan signs	Latin signs	contemporary Roman capitals
1			Aleph				Alpha			A
2			Beth				Beta			B
3			Gimel				Gamma			C
4			Daleth				Delta			D
5			He				Epsilon			E
6			Waw							F
										G
7			Zayin				Zeta			
8			Heth				Eta			H
9			Teth				Theta			
10			Yod				Iota			I
										J
11			Kaph				Kappa			K
12			Lamed				Lambda			L
13			Mem				Mu			M
14			Nun				Nu			N
15			Samekh				Xi			
16			Ayin				Omicron			O

No.			Name				Name					Latin
17			Pe			Π	Pi	Ρ	P			P
18			Sadhe					Μ				
19			Qoph					Q	Q			Q
20			Resh			Ρ	Rho	Ρ	R			R
21			Shin			Σ	Sigma		S			S
22			Taw			Τ	Tau	Τ	T			T
						Υ	Upsilon	Υ	V			V
												U
												W
						Φ	Phi	Χ				
						Χ	Chi	Φ	X			X
						Ψ	Psi	Υ				
						Ω	Omega					
							Digamma	Ϝ	Y			Y
									Z			Z

Table 1.

	Common Germanic	Thames	Vienna	Cod. Otho B X (10th Century)	Ruthwell
1	ᚠ f	ᚠ	ᚠ	ᚠ	ᚠ
2	ᚢ u	ᚢ	ᚢ	ᚢ	ᚢ
3	þ	þ	þ	þ	þ °
4	a	a °	ᚩ °	ᚩ °	ᚩ °
5	r	r ᶜ	r ᶜ	r ᶜ	r ᶜ
6	k	k ᶜ	k ᶜ	k	k ᶜ
7	g	X	X	X	X
8	w	ᚹ	ᚹ	ᚹ	ᚹ
9	h	ᚻ	ᚻ	ᚻ (ᚻ, I+)	ᚻ
10	n	+	+	+	+
11	i	—	—	—	—
12	j	+	°	° φ	[ʃ]
13	ë	ᛇ ë	ᛇ 'ih'	S Z 'co'	ᛋ ʃ
14	p	ᛈ x	ᛈ x	ᛈ ʒ	
15	z	ᛉ x	ᛉ x	ᛦ x	
16	s	ᛋ	ᛋ	ᛋ	ᛋ
17	t	ᛏ	ᛏ	ᛏ	ᛏ

No.	I	II	III	IV	V
18	ƀ	b	b	b	b
19	c	m	m	m	
20	m	n	m	m	m
21	l	d	l	l	l
22	ŋ	l	ŋ	ŋ	ŋ
23	o	m	d	œ	œ
24	t	œ	œ	d	d
25		a	a	a	a
26		æ	æ	æ	æ
27		y	ea	y	y
28		ĉa	y	îo	êa
29			ea	k	kᴵ
30			q	kᴵᴵ	
31			k	gᴵᴵ	kᴵᴵ
32				ŝt	
33				gᴵᴵ	gᴵᴵ

Table 2.

One can pinpoint the changes effected in the graph at each stage
by considering which 'elementary operations' are required to bring
about the transformation. By an <u>elementary operation</u> we mean one of
the following:

(a)   contraction of an edge or its inverse, expansion of an edge;

(b)   addition or deletion of an edge;

(c)   splitting of a vertex or edge;

(d)   identification of two vertices.

For example, the letter A in Figure 4(a) undergoes the following
elementary operation transformations: (i) ???; (ii) contract one
edge; (iii) contract three edges; (iv) expand one edge; (v) !!! .
An analysis of this type could conceivably help to settle certain
questions on the history of alphabets; palaeographers have various
theories on the origins of the runic alphabets. One could measure the
closeness of two alphabets by, for instance, the number of elementary
operations needed to transform the one into the other.

4.   <u>Pattern recognition</u>. An examination of the graphs of particular types of lettering and styles of writing would surely reveal
certain characteristics. (It might be of interest, in this regard,
to study children's handwriting.) This provides for a feasible approach to pattern recognition by computer. I shall give one example,
where the simplicity and beauty of a particular lettering is reflected,
at least to some extent, in its graphs -- Herbert Bayer's Universal
Type (see Figure 5).

We find that there are only nine graphs represented here, and
even then two of these are each represented by just one letter, namely
e and z. If these latter two were modified slightly so as to fit
into the classes {a, b, d, p, q} and {c, l, n, s, u}, respectively,
and if the dots on the i and j were omitted, there would be little
loss of legibility and we would be down to just a six graph alphabet
consisting of the first four stars, the loop graph, and the 'loop-
link' graph. This clearly indicates the simplicity of Bayer's
Universal Type.

<u>Figure</u> 5.

5. <u>The first graphographists</u>? The reader who has reached this point and still doubts the applicability of graph theory to palaeography has my sympathy. However, the natural association between alphabets and graphs is irrefutable in the light of the Chaldaic alphabet as illustrated by Geofroy Tory in a book entitled <u>Champ Fleury</u>, first published in 1529 (Table 3). The resemblance of many of these letters to graphs is so striking that I am tempted to conclude that even though Euler may have been the discoverer of the theory of graphs, Tory was aware of their existence long before. The following is an extract from <u>Champ Fleury</u>:

> I have followed the said Sigismund Fante, also in
> the names and figures of the <u>Chaldaic</u> letters, which
> are twenty-two in number & are also written from
> right to left, like the Hebrew & Arabic letters.
> Their names are as follows: <u>Aleph</u>, <u>Beth</u>, <u>Gimel</u>,
> <u>Daleth</u>, <u>He</u>, <u>Vau</u>, <u>Zain</u>, <u>Heth</u>, <u>Theth</u>, <u>Iod</u>, <u>Caph</u>,
> <u>Lamed</u>, <u>Mem</u>, <u>Nun</u>, <u>Samech</u>, <u>Hain</u>, <u>Pe</u>, <u>Zadi</u>, <u>Cof</u>, <u>Ress</u>,
> <u>Scin</u>, <u>Tau</u>. The said Fante says that the Hebrews
> used them in the time of Moses, when they were in
> the desert. His own words are as follows: <u>Questo</u>
> <u>soprascritto</u> <u>Alphabetto</u> <u>e</u> <u>Caldeo</u> <u>el</u> <u>quale</u> <u>usuano</u>,
> <u>li</u> <u>Hebrei</u> <u>nel</u> <u>tempo</u> <u>de</u> <u>Moyse</u> <u>nel</u> <u>deserto</u>. That
> is to say: 'This alphabet is the Chaldaic, which
> the Hebrews used in the time of Moses in the desert.'

6. <u>Calligraphic trail number</u>. We now introduce a new graph in-variant, the "calligraphic trail number", which is a measure of the difficulty of drawing the graph. It resembles the trail number -- the

Table 3.

minimum number of edge-disjoint trails required to cover the edges of
the graph -- except that it takes into account the problems one en-
counters in actually putting the graph down on paper. We always have

calligraphic trail number ≥ trail number.

The difference in the two invariants is arrived at in the following
way (slightly arbitrary, I admit, but serving its purpose quite well):

(i)   each vertex of degree two adds 1/2, since it represents
a change of direction.

Now imagine the trails drawn in, one by one;

(ii)   each time a trail starts at, passes through, or stops at an
already defined vertex, add 1/2; (remember that a vertex is defined
as the meeting-point of two or more lines having differing slopes, or
as the end of a line.)

(iii)   each time a trail starts on, touches, or stops on a previously drawn line (but not an already defined vertex) add 1/2.

It is here assumed that the embedding of the graph in the plane,
the choice of covering trails, and the order in which they are drawn
in, are such as to minimize contributions from (ii) and (iii); the sum
of the contributions from (i), (ii) and (iii) and the trail number is
called the <u>calligraphic trail number</u> and is denoted by $\mathscr{Y}(G)$. The
calligraphic trail number of a letter or symbol is that of its graph,
and is also denoted by $\mathscr{Y}(\cdot)$. Some examples should clarify the concept. In Figure 6 we indicate why $\mathscr{Y}(F) = 3$ and $\mathscr{Y}(H) = 4$.

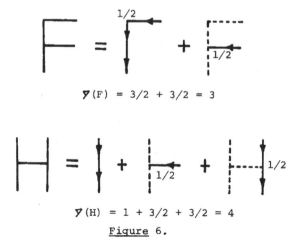

$$\mathscr{Y}(F) = 3/2 + 3/2 = 3$$

$$\mathscr{Y}(H) = 1 + 3/2 + 3/2 = 4$$

<u>Figure</u> 6.

It should be noted that $\gamma(K) = \gamma(X) = 2$, even though K is more difficult to draw than X; the definition of the calligraphic trail number of a letter or symbol can, of course, be modified to take into account its embedding in the plane.

It appears to be not quite trivial to find a formula for γ. (Compare the trail number.) However we have the following lower bound. For most small graphs this lower bound is, in fact, realized. Let n_k denote the number of vertices of degree k in G. Then

$$\gamma(G) \geq \begin{cases} 2 + n_0 + \frac{1}{2} \sum_{k \geq 2} (k-2)n_{2k}, \\ \qquad\qquad \text{if } n_2 = n_{2k+1} = 0, \text{ for all } k; \\[2ex] 1 + n_0 + \frac{1}{2}n_2 + \frac{1}{2} \sum_{k \geq 2} (k-2)n_{2k}, \\ \qquad\qquad \text{if } n_2 \neq 0, n_{2k+1} = 0, \text{ for all } k; \\[2ex] n_0 + \frac{1}{2}(n_1 + n_2) + n_3 + \frac{1}{2} \sum_{k \geq 2} (k-2)n_{2k} + kn_{2k+1}, \end{cases}$$

otherwise.

Finally, we note that $\gamma(H) = 4$, whereas $\gamma(\gamma) = 5/2$, which would seem to imply that γ is an easier letter to write than H. Why, then, did it disappear from use?

References

1. F. Harary, The Greek alphabet of "graph theory", Recent Progress in Combinatorics, (Proceedings Third Waterloo Conference on Combinatorics, 1968), Academic Press, New York, 1969, 13-20.

2. F. Harary, Typographs, Visibe Language, to appear.

3. A. Hill, Some problems from the visual arts, Proceedings of the International Conference on Combinatorial Mathematics, New York Academy of Sciences, 1970, 208-223.

THE NUMBER OF PARTIAL ORDER GRAPHS

Kim Ki-Hang Butler
Pembroke State University
Pembroke, NC 29372

We shall not present the details of proofs of various assertions since they are rather long and will be subsequently published in [4]. What we shall attempt to do instead is to give a rough indication of the proof and show how the results may be applied.

We find it convenient, although not essential, to phrase the argument in the terminology of Boolean relation matrix semigroup theory. Our chief tool in the study of this counting problem will be the Green's equivalence relations [6].

Let $V = \{v_1, \ldots, v_n\}$, so that $|V| = n > 0$. Let $D = (V, E)$ be a finite directed graph (digraph) with vertex set V and edge set E. Let $G_d(V)$ denote the set of all digraphs with vertex set V. For $D \in G_d(V)$ and $x \in V$ let $xD = \{y \in V\colon (x, y) \in E\}$. The non-empty set xD is called <u>rows</u> of D. The product of two graphs $D_1, D_2 \in G_d(V)$ is defined to be the graph D where $xD = (xD_1)D_2$ for $x \in V$. Geometrically this means that the row of D for the vertex x consists of all vertices which can be reached from x by an edge sequence of length 2 in which the first edge belongs to D_1 and the second to D_2, see [9]. Clearly, the product operation is associative. Hence $G_d(V)$ forms a semigroup. Since a graph is completely determined by either its adjacencies or its incidences, we may now rephrase this situation in the language of Boolean relation matrix semigroup theory.

By a <u>Boolean relation matrix</u> of order n is meant an $n \times n$ matrix over the Boolean algebra $B = \{0, 1\}$ of order 2. Let $B(n)$ denote the set of all such matrices. Then $B(n)$ forms a semigroup under the usual matrix multiplication. A digraph $D \in G_d(V)$ determines and is determined by an adjacency matrix $A(D) = (a_{ij}) \in B(n)$, where $a_{ij} = 1$ if there is an edge from v_i to v_j and $a_{ij} = 0$ otherwise (see Figure 1 for example). The correspondence $D \to A(D) = (a_{ij})$ is an isomorphism of the semigroup $G_d(V)$ onto the semigroup $B(n)$. In particular, we shall use the terms digraph and Boolean relation matrix interchangeably.

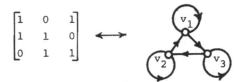

Figure 1.

In the remainder of this paper we shall work with special Boolean relation matrices. We shall need the following definitions. Let $V_n(B)$ denote the set of all row n-tuples over the Boolean algebra $B = \{0, 1\}$ of order 2. A nonempty subset W of $V_n(B)$ is said to be <u>linearly independent</u> if the zero n-tuple $z = (0, \ldots, 0)$ is not in W and no member of W is a sum of other members of W. A <u>subspace</u> of $V_n(B)$ is a nonempty subset closed under addition. A linearly independent subset of $V_n(B)$ that generates a subspace W of $V_n(B)$ under addition is called a <u>basis</u> of W. To each $A \in B(n)$ we associate the following sets: (i) the <u>row space</u> $R(A) = \{xA: x \in V_n(B)\}$, (ii) the <u>column space</u> $C(A) = \{Ax: x \in V^n(B)\}$, where $V^n(B) = \{y^t: y \in V_n(B)\}$ and y^t denotes the transpose of y. For $A \in B(n)$ the basis of the subspace of $V_n(B)$ generated by the nonzero rows of A is called the <u>row basis</u> of A and its cardinality is called the <u>row rank</u> of A. The definitions of column basis and column rank are given in a similar manner. Let $\rho_r(A)$ and $\rho_c(A)$ denote the row and column rank of A, respectively. It is known that every finite row (column) space has a unique basis and that if $A \in B(n)$ then we need not have $\rho_r(A) = \rho_c(A)$, see [2]. A matrix $A \in B(n)$ is called <u>regular</u> if $A = AXA$ for some X in $B(n)$. We shall call a matrix $A \in B(n)$ <u>nonsingular</u> if A is regular and $\rho_r(A) = n = \rho_c(A)$. A matrix $A \in B(n)$ is called <u>idempotent</u> if $A^2 = A$. If $A \in B(n)$ is both nonsingular and idempotent, then we say that A is <u>nonsingular idempotent</u>.

Let us see how partial order graphs are connected with Boolean relation matrices. By a <u>partial order relation</u> on V is meant a reflexive, antisymmetric and transitive relation on V. The corresponding graph is therefore transitive with loops and there is at most a single edge connecting two vertices. In other words, D is called a <u>partial order graph</u>, alias <u>pograph</u>, on V if the relation $\leq$ defined on V by the rule that x ≤ y iff there is an edge from x to y in D is a partial order relation. Let $G_p(V)$ denote the set of all pographs on V. For simplicity, we denote $P(n) = |G_p(V)|$. From elementary matrix theory the following technical lemma is easily obtained.

LEMMA 1. If $A(D) = (a_{ij}) \in B(n)$ is an adjacency matrix of a pograph D, then A has the following properties:

(1) $a_{ii} = 1$ for $i = 1, \ldots, n$ (reflexive),

(2) $a_{ij} = 1$ for $i \neq j \Rightarrow a_{ji} = 0$ (antisymmetric),

(3) $a_{ij} = 1$ and $a_{jk} = 1 \Rightarrow a_{ik} = 1$ (transitive).

It follows that (i) A(D) is in dominant diagonal form, that is,

$$A(D) = \begin{bmatrix} 1 & * & * & \ldots & * \\ * & 1 & * & \ldots & * \\ * & * & 1 & \ldots & * \\ & & \ldots\ldots & & \\ * & * & * & \ldots & 1 \end{bmatrix}_{n \times n} ,$$

where the asterisks denote 0 or 1; (ii) $\rho_r(A(D)) = n = \rho_c(A(D))$; (iii) $(A(D))^2 = A(D)$; and (iv) $A(D) = A(D) \times A(D)$ for some $X \in B(n)$.

COROLLARY 2. A matrix $A(D) \in B(n)$ is an adjacency matrix of a pograph D if and only if $A(D) \in E(n)$.

It follows from Corollary 2 that there exists a natural one-to-one correspondence between E(n) and P(V), the set of all partially ordered sets (posets) defined on V. Thus, the enumeration of pographs with vertex set containing n elements is equivalent to the enumeration of nonsingular idempotent matrices of B(n).

THEOREM 3. For $1 \leq n \leq 7$, P(n) is given by the following table:

n	1	2	3	4	5	6	7
P(n)	1	3	19	219	4231	130023	6129859

Proof. The proof follows from Corollary 2 and enumeration of posets found in [1, p. 4], [7], and [10].

We now turn our attention to the enumeration of $P(n)$ for $n > 7$. In order to separate the set $E(n)$ into subsets, we shall need the following idea. Two elements of an arbitrary semigroup S are said to be $\mathcal{L}$-equivalent if they generate the same principal left ideal of S. $\mathcal{R}$-equivalence is defined dually. The join of equivalence relations $\mathcal{L}$ and $\mathcal{R}$ is denoted by $\mathcal{D}$ and their intersection by $\mathcal{H}$. These equivalence relations are called Green's relations on S [6, Chapter 2]. It has been shown [6] that in any $\mathcal{D}$-class D of S, either every element is regular or else every element is irregular. If the matrices of the $\mathcal{D}$-class D of $B(n)$ are nonsingular then D is called a nonsingular $\mathcal{D}$-class. Let S_n denote the group of permutation matrices of order n. The next two lemmas are evident from [2, Theorem 8].

LEMMA 4. Let A and B be nonsingular matrices of $B(n)$. Then the following statements are equivalent:

(1) $A \mathcal{L} B$ $(A \mathcal{R} B)$,

(2) $R(A) = R(B)$ $(C(A) = C(B))$,

(3) $A = PB$ $(A = BP)$ for some $P \in S_n$.

LEMMA 5. Let A and B be nonsingular matrices of $B(n)$. Then

(1) $A \mathcal{D} B$ if and only if there exist $P, Q \in S_n$ such that $A = PBQ$,

(2) $A \mathcal{H} B$ if and only if there exist $P, Q \in S_n$ such that $A = PB = BQ$.

THEOREM 6. Each nonsingular $\mathcal{L}$-class of $B(n)$ contains exactly one nonsingular idempotent matrix. A similar result holds for the $\mathcal{R}$-classes.

Proof. Let $A = (a_{ij})$, $B = (b_{ij}) \in E(n)$, where $A \mathcal{L} B$. Then by Lemma 4(3), there exists $P \in S_n$ such that $A = PB$. By Lemma 1(1), $a_{ii} = 1 = b_{ii}$ for $i = 1, \ldots, n$. Hence $P = I_n$, the $n \times n$ identity matrix, and so $A = B$. A similar proof holds for the $\mathcal{R}$-classes.

It is a consequence of this theorem that the enumeration of $E(n)$ is equivalent to the enumeration of nonsingular $\mathcal{L}$-classes of $B(n)$.

In order to enumerate the nonsingular $\mathcal{L}$-classes of $B(n)$, we shall need the following definition. Two matrices A and B of $E(n)$ are isomorphic iff there exists a unique permutation matrix $P \in S_n$ such that $A = PBP^{-1}$, where P^{-1} denotes the inverse of P.

THEOREM 7. Two matrices A and B of $E(n)$ are isomorphic if and only if $A \mathcal{D} B$.

Proof. Let $A, B \in E(n)$. Then by Lemma 5 (1), $A \mathcal{D} B$ iff there exist $P, Q \in S_n$ such that $A = PBQ$. Hence we need only to show that Q is an inverse of P. Since $A, B \in E(n)$, $A = PBQ = (PBQ)(PBQ) = PBRBQ$, which implies that $B = BRB$, where $R = QP$. Similarly, $B = BRB = (BRB)(BRB) = B(RBR)B$ which implies that

$$
RBR = \begin{cases} R & \text{if } B = I_n, \\ \\ B & \text{otherwise.} \end{cases}
$$

where I_n is an $n \times n$ identity matrix. From this we obtain $RBR = B$, which in turn implies that $R = I_n = QP$. Hence Q is an inverse of P.

Let $E(n, r)$ denote the subset of $E(n)$ containing matrices with exactly r 1's off the main diagonal. Clearly $E(n, r)$ is nonempty only when $0 \le r \le n(n - 1)/2$. To avoid notational complications, let us assume $p = n(n - 1)/2$. Then it follows from the definition that

$$
P(n) = \sum_{r=0}^{p} |E(n, r)|.
$$

COROLLARY 8. Let $A, B \in E(n)$. Then $A \mathcal{D} B$ only if $A, B \in E(n, r)$.

We must point out that the converse of the corollary does not hold.

The isomorphism and nonisomorphism of two matrices of $E(n, r)$ can be tested most simply by drawing their Hasse diagrams; see Fig.2.

$$A = \begin{bmatrix} 1 & 0 & 0 \\ 0 & 1 & 0 \\ 1 & 1 & 1 \end{bmatrix} \longleftrightarrow$$

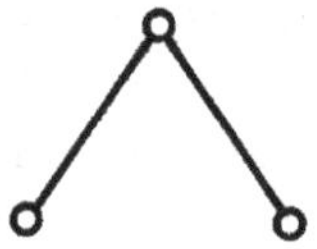

$$B = \begin{bmatrix} 1 & 0 & 0 \\ 1 & 1 & 0 \\ 1 & 0 & 1 \end{bmatrix} \longleftrightarrow$$

Figure 2.

Let $E^*(n)$ denote the set of all nonisomorphic matrices of $E(n)$, and let $|E^*(n)| = e_n$. The following theorem was obtained first by Rose and Sasaki [1] and then independently discovered by Wright in [10].

THEOREM 9. For $0 \le n \le 7$, e_n is given by the following table:

n	0	1	2	3	4	5	6	7
e_n	1	1	2	5	16	63	318	2045

Let $E^*(n, r)$ be the set of all nonisomorphic matrices of $E(n, r)$, and let $|E^*(n, r)| = m_r$. Suppose $A \in E^*(n, r)$. If we permute the rows or columns of A then we obtain a matrix B such that $A \mathrel{\mathcal{S}} B$. Obviously, every matrix of $E^*(n, r)$ generates a $\mathcal{S}$-class and $E^*(n, r)$ generates exactly m_r distinct nonsigular $\mathcal{S}$-classes of $B(n)$. Such $\mathcal{S}$-classes will be denoted by $D(n, r, q)$ $(q = 1, \ldots, m_r)$. We define

$$D(n, r) = \bigcup_{q=1}^{m_r} \{D(n, r, q)\}.$$

Clearly, $|E^*(n, r)| = m_r = |D(n, r)|$.

The following theorem will be useful in enumerating the elements of $E(n)$; the theorem was proved by using techniques of lattice theory.

THEOREM 10. For $p = n(n - 1)/2$, the following table provides values of m_r for the given seven values of r:

r	0	1	2	3	p - 2	p - 1	p
m_r	1	1	3	7	$n(n-2)/2$	$n-1$	1

We now proceed to enumerate the elements of $E(n)$. We shall need the following known results. The $\mathcal{H}$-classes contained in the same $\mathcal{D}$-class have the same cardinal number [6, Theorem 2.3]. Let $h(r, q)$ denote the cardinality of an $\mathcal{H}$-class contained in $D(n, r, q)$.

THEOREM 11. The number $P(n)$ is given by the formula

$$P(n) = \sum_{r=0}^{p} \left(\sum_{q=1}^{m_r} n!/h(r,q) \right).$$

Proof: Let $A \in D(n, r, q)$, where r and q are fixed. If R_A denotes the $\mathcal{R}$-class containing A, then by Lemma 4 the elements of R_A are all obtained by permuting the columns of A, and so $|R_A| = n!$. This implies that the number of possible $\mathcal{L}$-classes contained in a $\mathcal{D}$-class $D(n, r, q)$ is $n!$. But if there exists $B \in R_A$ such that $A \mathcal{L} B$, then $A \mathcal{H} B$. Hence in order to obtain the exact number of $\mathcal{L}$-classes contained in $D(n, r, q)$, we must divide $n!$ by $h(r, q)$. We sum over all $\mathcal{L}$-classes contained in the various nonsingular $\mathcal{D}$-classes of $B(n)$, to get the desired result.

The above theorem can be formulated in terms of groups. It is known that any $\mathcal{H}$-class containing an idempotent is a group [6, Theorem 2.16], and that two such $\mathcal{H}$-classes in the same $\mathcal{D}$-class are isomorphic [6, Theorem 2.20]. Now by Theorem 6, every $\mathcal{L}$ ($\mathcal{R}$)-class contained in $D(n, r, q)$ contains a group (i.e., $\mathcal{H}$-class containing a nonsingular idempotent matrix). Such groups will be called nonsingular $\mathcal{D}$-equivalent groups. In other words, if $H_i \in D(n, r, i)$ is a group and $H_j \in D(n, r, j)$ is a group, where $i \neq j$, then H_i and H_j are called nonsingular $\mathcal{D}$-inequivalent groups. There are exactly e nonsingular $\mathcal{D}$-inequivalent groups contained in $B(n)$. We may rephrase Theorem 11 as follows.

THEOREM 12. If H_i (i = 1, ..., e_n) give all the different nonsingular $\mathcal{D}$-inequivalent groups of $B(n)$, then

$$P(n) = \sum_{i=1}^{e_n} n!/o(H_i),$$

where $o(H_i)$ denotes the order of the group H_i.

In view of Theorems 11 and 12, we still have to enumerate $h(r, q)$, for all r and q. Determining the values of $h(r, q)$ is generally a very hard combinatorial problem. However, we have results for $r = 0, 1, 2, 3, p-2, p-1$, and p. We abbreviate $|E(n, r)|$ by $P(n, r)$. Then it follows from the definition that

$$P(n) = \sum_{r=0}^{p} P(n, r) = \sum_{r=0}^{p} |E(n, r)| = |E(n)|.$$

THEOREM 13. For $p = n(n - 1)/2$, the following table provides values of $P(n, r)$ for the given seven values of r:

r	$P(n, r)$
0	1
1	$n(n - 1)$
2	$3n(n - 1)(n - 2)/2$
3	$n(n - 1)(n - 2)(3n^3 - 33n^2 + 124n - 153)/3$
$p - 2$	$n!(2n^2 - 6n + 1)/4$
$p - 1$	$n!(n - 1)/2$
p	$n!$

References

1. G. Birkhoff, Lattice Theory, 3rd Ed., Amer. Math. Soc., Providence, R. I., 1967.

2. K. K.-H. Butler, Binary relations, Recent Trends in Graph Theory, Lecture Notes in Mathematics, No. 186, Springer-Verlag, Berlin, (1971), 25-47.

3. _____, On (0, 1)-matrix semigroups, Semigroup Forum 3 (1971), 74-79.

4. _____, The number of partially ordered sets, to appear.

5. _____, Canonical bijection between $\mathscr{D}$-classes of (0, 1)-matrix semigroups, *Periodica Mathematica*, to appear.

6. A. H. Clifford and G. Preston, *The Algebraic Theory of Semigroups*, Vol. 1, Amer. Math. Soc., Providence, R. I., 1961.

7. J. W. Evans, F. Harary, and M. S. Lynn, On the computer enumeration of finite topologies, *Comm. Assoc. Comp. Mach.* 10 (1967), 295-298.

8. F. Harary, *Graph Theory*, Addison Wesley, Reading, Mass., 1969.

9. R. Plemmons, Graphs associated with a group, *Proc. Amer. Math. Soc.* 25 (1970), 273-276.

10. J. A. Wright, *Cycle indices of certain classes of quasiorder types or topologies*, Doctoral Thesis, University of Rochester, 1972.

ESTIMATING THE CONNECTIVITY OF A GRAPH

Michael Capobianco
Notre Dame College of St. John's University
Staten Island, NY 10301

1. Introduction. In a previous paper [1], we introduced the
notion of statistical inference in graphs and digraphs. This idea
gives rise to a staggering variety of problems including the recon-
struction conjecture as a special case. In this paper we deal with
just one of these many problems. The work presented here is certainly
only a small first step and there is much more to be done.

The inference problem with which we are concerned can be stated
abstractly as follows: Given a collection of subgraphs of a connected
labeled graph G, derive an estimate of the connectivity of G.
Statistically speaking we regard the unknown graph G as a structured
population and the subgraphs as samples from this population. One can
imagine many applications. The graph G could represent a social
group structure, a transportation or communication network, a pattern
of ecological relationships, etc. Depending on the application, the
samples will be of different types, e.g. induced subgraphs of two
points, star subgraphs, induced subgraphs of more than two points,
samples related to paths between two points. Each different type
gives rise to a different approach to the problem. We will be concerned mainly with star subgraphs, although we will also discuss induced subgraphs of two points to some extent.

2. Star subgraph samples. A star subgraph sample of size n is
obtained by choosing n points of G at random and observing the
star subgraphs of which these points are the centers. It is assumed
that the labels of the noncentral points are also observable. For
example, suppose G is the graph diagramed below.

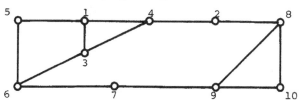

(All <u>we</u> know about G is that it is connected and has 10 points.)

Now suppose that n = 3, and that the randomly chosen points turned out to be those labeled 1, 2, 3. Then the samples we would observe would be

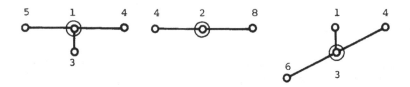

We could put these together to get to get the following graph.

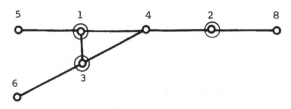

(The original centers have been encircled.)

An idea which comes to mind immediately for estimating connectivity using star subgraphs is to make use of the theorem of Whitney [5]: $\kappa \le \lambda \le \delta$, where δ is the minimum degree in the graph, λ is the line connectivity, and κ is the (point) connectivity. We will indeed use this in a natural way by taking the minimum degree of the centers in the sample, $\hat{\delta}$, as an estimate of δ. What we will be estimating, of course, is only an upper bound on κ. We will look into what could be done to improve this later. The remainder of this section will be devoted to an investigation of the properties of the estimator $\hat{\delta}$.

Our first task will be to find the probability distribution of $\hat{\delta}$. To this end, let G have p_i points of degree i, i = 1, 2, 3, ..., p-1. The set of numbers $p_1, p_2, p_3, ..., p_{p-1}$ is what Chinn calls the <u>frequency partition</u> of G [3]. Then $\sum_{i=1}^{p-1} p_i = p$ and $\sum_{i=1}^{p-1} i\, p_i = 2q$, where q is the number of lines in G. Now of course $\sum_{i=1}^{\delta-1} p_i = 0$ and $P(\hat{\delta} = k) = 0$ for all k < δ. Furthermore, let j be the smallest integer such that n (the sample size) is greater than the total number of points of degree greater than j, i.e.

$$n \le \sum_{i=j}^{p-1} p_i \; ,$$

but

$$n > \sum_{i=j+1}^{p-1} p_i \; ;$$

then $P(\hat{\delta} = k) = 0$ for all $k > j$. For $\delta \le k < j$ we have

$$P(\hat{\delta} = k) = \frac{\binom{\sum_{i=k}^{p-1} p_i}{n} - \binom{\sum_{i=k+1}^{p-1} p_i}{n}}{\binom{p}{n}} . \qquad (1)$$

To see this it suffices to realize that

$$P(\hat{\delta} \ge k) = \frac{\binom{\sum_{i=k}^{p-1} p_i}{n}}{\binom{p}{n}}$$

and that

$$P(\hat{\delta} > k) = \frac{\binom{\sum_{i=k+1}^{p-1} p_i}{n}}{\binom{p}{n}} .$$

Now if $k = j$ then $P(\hat{\delta} > k) = 0$ and therefore

$$P(\delta = k) = \frac{\binom{\sum\limits_{i=k}^{p-1} P_i}{n}}{\binom{p}{n}} .$$

Note that if we adopt the usual convention that $\binom{m}{r} = 0$ if $r > m$ then (1) holds for all k.

Next we compute the expected value of $\hat{\delta}$. The expression $\sum\limits_{i=k}^{p-1} P_i$ will be denoted by Σ_k.

$$E(\hat{\delta}) = \sum_{k=1}^{p-1} kP(\hat{\delta} = k) = \sum_{k=\delta}^{j-1} \frac{k\left[\binom{\Sigma_k}{n} - \binom{\Sigma_{k+1}}{n}\right]}{\binom{p}{n}} + \frac{j\binom{\Sigma_j}{n}}{\binom{p}{n}}$$

$$= \frac{1}{\binom{p}{n}}\left(\delta\binom{p}{n} - \delta\binom{P-P_\delta}{n} + (\delta+1)\binom{P-P_\delta}{n} - (\delta+1)\binom{P-P_\delta-P_{\delta+1}}{n} + \cdots\right.$$

$$\left. \cdots + (j-1)\binom{\Sigma_{j-1}}{n} - (j-1)\binom{\Sigma_j}{n} + j\binom{\Sigma_j}{n}\right)$$

$$= \frac{1}{\binom{p}{n}}\left[\delta\binom{p}{n} + \binom{P-P_\delta}{n} + \binom{P-P_\delta-P_{\delta+1}}{n} + \cdots + \binom{\Sigma_j}{n}\right]$$

$$= \delta + \frac{1}{\binom{p}{n}} \sum_{i=\delta}^{j-1} \binom{P-\sum\limits_{t=\delta}^{i} P_t}{n} = \delta + \frac{1}{\binom{p}{n}} \sum_{i=\delta+1}^{j} \binom{\Sigma_i}{n} .$$

In statistics, the _bias_ of an estimator of a parameter θ is defined as $E(t) - \theta$ where t is the estimator. If this is zero, then t is said to be _unbiased_. This is a desirable property since it means that "on the average" the estimator equals the parameter it is intended to estimate.

As can be seen from the above result, $\hat{\delta}$ is _not_ an unbiased estimator of δ. In fact, its bias is non-negative, i.e. it is "biased high". This is not at all suprising since $\delta \le \hat{\delta}$. Note, however, that the bias approaches zero as n approaches p, i.e., $\hat{\delta}$ is "asymptotically unbiased".

We propose a procedure to help improve this estimator. Since it does not appear to be a simple matter to convert $\hat{\delta}$ into an unbiased estimator, we suggest estimating the bias from the sample and then subtracting this estimate from $\hat{\delta}$. We shall denote this estimator by $\hat{\delta}*$. We then need estimators of p_δ, $p_{\delta+1}$, $p_{\delta+2}$, $\ldots$, p_{j-1}. These will be obtained by the well-known method of maximum likelihood. This method chooses the estimator which maximizes the probability of the observed sample. This can be expressed as

$$f(x_1, x_2, \ldots, x_n) = \prod_{i=1}^{n} \frac{p_{x_i}}{p} = \frac{\displaystyle\prod_{i=1}^{n} p_{x_i}}{p^n} \ ,$$

where x_i is the degree of the ith center in the sample. To maximize this with respect to the p_i's we take its logarithm and set the partial derivatives equal to zero. Thus,

$$\log f = -n \log p + \sum_{i=1}^{n} \log p_{x_i} \ ,$$

$$\frac{\partial \log f}{\partial p_k} = -\frac{n}{p} \frac{\partial p}{\partial p_k} + \frac{x(k)}{p_k} \ ,$$

where $x(k)$ is the number of centers in the sample having degree k. Hence,

$$\frac{n}{p} = \frac{x(k)}{\hat{p}_k}$$

or

$$\hat{p}_k = \frac{x(k)}{n} p \ ,$$

quite a sensible result.

To illustrate this procedure consider the graph below.

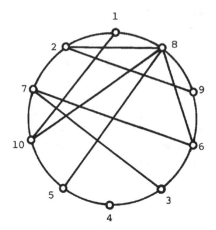

We used a sample size of 5. Hence, 5 random numbers were chosen from a standard table. The results were 5, 9, 2, 1, and 7. Therefore, our samples (put together) appear as drawn below.

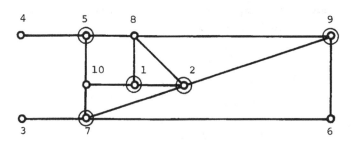

From this we see that $\hat{\delta} = 3$, $\hat{P}_3 = 6$, $\hat{P}_4 = 4$, and all other $\hat{P}_i$'s are 0. Hence the estimated bias is

$$\frac{1}{\binom{10}{5}}\binom{4}{5} = 0 .$$

Therefore, $\hat{\delta}^* = 3$, so that $\hat{\kappa} \le 3$. Note that the actual δ is 2; κ is also 2. Note also that j is estimated as 4.

It is interesting to compute the variance of $\hat{\delta}$. For this we use the standard formula

$$\text{var}(\hat{\delta}) = E(\hat{\delta}^2) - [E(\hat{\delta})]^2 .$$

We first compute $E(\hat{\delta}^2)$:

$$E(\hat{\delta}^2) = \sum_{k=1}^{p-1} k^2 p(\hat{\delta} = k) = \sum_{k=\delta}^{j-1} \frac{k^2\left[\binom{\Sigma_k}{n} - \binom{\Sigma_{k+1}}{n}\right]}{\binom{p}{n}} + \frac{j^2\binom{\Sigma_j}{n}}{\binom{p}{n}}.$$

Now if $j = \delta$ then $E(\hat{\delta}^2) = \delta^2$. Assuming $j > \delta$ we have

$$E(\hat{\delta}^2) = \frac{1}{\binom{p}{n}}\left[\delta^2\binom{p}{n} - \delta^2\binom{\Sigma_{\delta+1}}{n} + (\delta+1)^2\binom{\Sigma_{\delta+1}}{n} - (\delta+1)^2\binom{\Sigma_{\delta+2}}{n} + \dots \right.$$

$$\left. \dots + (j-1)^2\binom{\Sigma_{j-1}}{n} - (j-1)^2\binom{\Sigma_j}{n} + j^2\binom{\Sigma_j}{n}\right]$$

$$= \frac{1}{\binom{p}{n}}\left[\delta^2\binom{p}{n} + (2\delta+1)\binom{\Sigma_{\delta+1}}{n} + (2\delta+3)\binom{\Sigma_{\delta+2}}{n} + \dots \right.$$

$$\left. \dots + (2j-1)\binom{\Sigma_j}{n}\right]$$

$$= \delta^2 + \frac{1}{\binom{p}{n}}\left[\sum_{i=\delta+1}^{j} (2i-1)\binom{\Sigma_j}{n}\right].$$

Therefore,

$$\text{var}(\hat{\delta}) = \frac{1}{\binom{p}{n}}\sum_{i=\delta+1}^{j} (2(i-\delta)-1)\binom{\Sigma_i}{n} - \frac{1}{\left[\binom{p}{n}\right]^2}\left(\sum_{i=\delta+1}^{j}\binom{\Sigma_i}{n}\right)^2.$$

Note that as $n \to p$, $\text{var}(\delta) \to 0$. It is a theorem of mathematical statistics that an asymptotically unbiased estimator having the above property is consistent, i.e., it converges in probability to the parameter it is estimating.

In this case, in fact, the variance is 0 provided $n > \Sigma_{\delta+1}$, so that $\hat{\delta}$ is always equal to δ under these conditions. Of course, this is obvious from the start because the condition $n > \Sigma_{\delta+1}$ means that our sample consists of more center points than there are points of degree greater than δ in G.

3. <u>Samples</u> <u>of</u> <u>subgraphs</u> <u>induced</u> <u>by</u> <u>two</u> <u>points</u>. The estimator used with this type of sampling is based on the result of Harary [4] to the effect that if $q \geq p-1$ then $k \leq \left[\frac{2q}{p}\right]$. Clearly, each subgraph in our sample will be either K_2 or $\overline{K}_2$. The idea is to give an estimate L of the number $\left[\frac{2q}{p}\right]$ by estimating q using the number of K_2's in a sample of n pairs of points. This problem has been previously discussed by us [1]. The distribution of L is hypergeometric and

$$E(L) = \frac{nq}{\binom{p}{2}} ,$$

$$var(L) = \frac{nq\left(\binom{p}{2}-q\right)\left(\binom{p}{2}-n\right)}{\left[\binom{p}{2}\right]^2\left(\binom{p}{2}-1\right)} .$$

This again yields a biased estimator, in fact, one that is neither asymptotically unbiased nor consistent. ($E(L) = q$ only when n <u>equals</u> $\binom{p}{2}$.) However, the bias is easily corrected in this case. One need only use the estimator

$$L^* = \frac{\binom{p}{2}}{n} L .$$

Then

$$E(L^*) = q$$

and

$$var(L^*) = \frac{\left[\binom{p}{2}\right]^2}{n^2} var(L) = \frac{q\left(\binom{p}{2}-q\right)\left(\binom{p}{2}-n\right)}{n\left(\binom{p}{2}-1\right)} .$$

This is not consistent, since $var(L^*) = 0$ only when n <u>equals</u> $\binom{p}{2}$.

Hence our estimator of maximum connectivity ought to be taken as

$$M = \left[\frac{2L^*}{p}\right] = \left[\frac{2\binom{p}{2}L}{np}\right] = \left[\frac{(p-1)L}{n}\right].$$

It is clear that M is an estimate of the mean degree of G so that "on the average" $M \geq \hat{\delta}$. Why, then, ever use M? The answer to this question must be found in the practical aspects of the application involved. Subgraphs induced by two points may be the only kind of samples one can take. They may be less expensive than star samples. These considerations must be weighed against the relative importance of overestimating and underestimating connectivity. It is a statistical decision problem!

We illustrate the use of M with the same graph as in the previous section. Our sample consists of 23 pairs chosen at random. The results (arranged lexicographically) were (1,2) (1,5) (1,8) (2,3) (2,4) (2,5) (2,6) (2,7) (2,8) (2,9) (3,4) (3,5) (3,8) (3,9) (3,10) (4,7) (5,6) (5,8) (6,7) (6,8) (6,9) (7,10) (9,10). As can be seen, $L = 11$. Hence, $M = \left[\frac{99}{23}\right] = 4$.

4. Conclusion. Although the results above seem to yield good estimators of an upper bound on connectivity, they can easily give rather poor estimates of κ itself. One possible improvement could be effected by using the theorem of Harary and Chartrand [2] that $\delta \geq \frac{p-2+n}{2}$ for some n such that $1 \leq n \leq p-1$ implies $\kappa \geq n$. This could give an estimate of a lower bound on κ by using δ^* in place of δ in the above inequality. Of course, the usefulness of this is limited to cases in which δ^* is rather large, at least $\frac{1}{2}p$.

Further approaches to this problem based on testing the hypothesis that $\kappa = 1$ will hopefully be the subject of a future paper.

References

1. M. Capobianco, Statistical inference in finite populations having structure, Trans. New York Acad. Sci. 32 (1970), 401-143.

2. G. Chartrand and F. Harary, Graphs with prescribed connectivities, _Theory_ _of_ _Graphs_, (P. Erdös and G. Katona, Eds.) Akademiai Kiado, Budapest, (1968), 61-63.

3. P. Zweig Chinn, The frequency partition of a graph, _Recent_ _Trends_ _in_ _Graph_ _Theory_, (M. Capobianco, J. Frechen, M. Krolik, Eds.), Springer-Verlag, (1971).

4. F. Harary, The maximum connectivity of a graph, _Proc_. _Nat_. _Acad_. _Sci_. _USA_ 48 (1962), 1142-1146.

5. H. Whitney, Congruent graphs and the connectivity of graphs, _Amer_. _J_. _Math_. 54 (1932), 150-168.

ON PROBLEMS OF MOSER AND HANSON

P. Erdős
Imperial College
London, England

and

S. Shelah
Hebrew University
Jeruselem, Israel

The following problem is due to L. Moser: Let $A_1, \ldots, A_n$ be any n sets. Take the largest subfamily $A_{i_1}, \ldots, A_{i_r}$ which is union-free, i.e.,

$$A_{i_{j_1}} \cup A_{i_{j_2}} \neq A_{i_{j_3}}, \quad 1 \leq j_1 \leq r, \quad 1 \leq j_2 \leq r, \quad 1 \leq j_3 \leq r,$$

for every triple of distinct sets $A_{j_1}, A_{j_2}, A_{j_3}$. Put $f(n) = \min r$, where the minimum is taken over all families of n distinct sets. Determine or estimate $f(n)$. Riddel showed $f(n) > c\sqrt{n}$ and Erdős and Komlós [1] showed

$$\sqrt{n} \leq f(n) \leq 2\sqrt{2}\sqrt{n}. \tag{1}$$

We now show

$$\sqrt{2n} - 1 < f(n) < 2\sqrt{n} + 1 \tag{2}$$

and we conjecture that $f(n) = (2 + o(1))\sqrt{n}$.

Consider now the largest subfamily $A_{i_1}, \ldots, A_{i_r}$ so that no four distinct sets satisfy

$$A_{i_{j_1}} \cup A_{i_{j_2}} = A_{i_{j_3}}, \quad A_{i_{j_1}} \cap A_{i_{j_2}} = A_{i_{j_4}}. \tag{3}$$

Put $F(n) = \min r$, where the minimum is taken over all families of distinct sets $A_1, \ldots, A_n$. We prove

(4)
$$F(n) \le \frac{3}{2} n^{2/3}.$$

Probably $F(n) > c_2 n^{2/3}$, and in fact it seems likely that $F(n)/n^{2/3}$ tends to a limit, but we have not been able to show this.

Hanson posed the following problem: Let $|Q| = n$, with $g(n)$ the smallest integer so that the subsets of Q can be split into $g(n)$ classes where each of the classes is union free. Hanson proved

(5)
$$c_3 \sqrt{n} < g(n) \le \frac{n}{2} + 2$$

and he conjectured that the upper bound is substantially correct. We prove

(6)
$$g(n) > \frac{n}{4}.$$

Let $G(n)$ be the smallest integer so that the subsets of Q can be split into $G(n)$ classes so that no class contains four distinct sets A_1, A_2, A_3, A_4 satisfying (3). We prove

(7)
$$c_4 \sqrt{n} < G(n) < c_5 \sqrt{n}.$$

Probably $\lim_{n \to \infty} G(n)/n^{1/2}$ exists.

Now we prove (2). We use a slight improvement of the method of [1] to prove the upper bound. Let t be the least integer for which $[t^2/4] > n$. Our A's are the $[t^2/4]$ set of integers $A_{i,j} = \{x: i \le x \le j\}$, $1 \le i \le t/2 < j \le t$. We show that the largest union-free subfamily of the A's has at most t elements. To see this let A_{i_r,j_r}, $1 \le r \le \ell$, be a union-free subfamily of the A's. An endpoint i_r (or j_r) is called good if there is no other A_{i_s,j_s} of our family with $i_r = i_s$ and $j_s < j_r$ (or $j_r = j_s$, $i_s > i_r$). Clearly at least one endpoint of A_{i_r,j_r} must be good, for otherwise A_{i_r,j_r} would be the union of two A's of our family. But an integer can be a good endpoint of at most one A_{i_r,j_r}, which shows $\ell \le t$ and our assertion is proved. Now clearly

$$f(n) \le f([t^2/4]) \le t$$

or $f(n) \leq 2\sqrt{n} + 1$. which proves the upper bound of (2).

We now prove the lower bound. Let $\{A_1, \ldots, A_n\}$ be any family of n distinct sets. We define a union-free subfamily $\{A_{i_1}, \ldots, A_{i_r}\}$ as follows. A_{i_1} is any minimal A, i.e., contains no other as a proper subset. Suppose $A_{i_1}, \ldots, A_{i_s}$ have already been defined. Then $A_{i_{s+1}}$ is chosen to be a minimal member of $\{A_1, \ldots, A_n\} \setminus \{A_{i_1}, \ldots, A_{i_s}\}$ which is not the union of two distinct members of $\{A_{i_1}, \ldots, A_{i_s}\}$. There is clearly a choice for $A_{i_{s+1}}$ if $n - s > \binom{s}{2}$. This process therefore defines a subfamily $\{A_{i_1}, \ldots, A_{i_r}\}$ of r sets, where $r + \binom{r}{2} \geq n$, i.e., $r \geq \sqrt{2n} - 1$. To complete the proof it only remains to show that the family $\{A_{i_1}, \ldots, A_{i_r}\}$ is union-free.

Assume

$$A_{i_j} \cup A_{i_k} = A_{i_\ell} \quad (j \neq \ell, \; k \neq \ell). \tag{8}$$

We cannot have $\ell > j$ and $\ell > k$ by the construction. So we can assume $k > \ell$. But this is also impossible, since A_{i_ℓ} was chosen ad a minimal member of $\{A_1, \ldots, A_n\} \setminus \{A_{i_1}, \ldots, A_{i_{\ell-1}}\}$. Hence (8) cannot hold and the proof of (2) is complete.

It is not difficult to improve the lower bound of (2) slightly to show that $f(n) > (1+c)\sqrt{2n}$. However, we cannot show that $f(n) = (2 + o(1))\sqrt{n}$.

To prove (4) we use an idea due to Folkman. Let t be the least integer for which $t^3 \geq n$. Consider the t^3 sets $A_{i,j} = \{n: i \leq n \leq j\}$, $(1 \leq i \leq t < j \leq t^2 + t)$. Thus the sets $A_{i,j}$ correspond to the edges of the complete bipartite graph $K(t, t^2)$. A simple argument shows that every subgraph of $K(t, t^2)$ having $t^2 + \binom{t}{2} + 1$ edges contains a rectangle; this is false for $t^2 + \binom{t}{2}$ edges. A rectangle corresponds to four distinct sets A satisfying (3). Thus

$$F(n) \leq F(t^3) \leq t^2 + \binom{t}{2},$$

which proves (4).

Instead of (3) we could consider other systems of equations with sets as unknowns, but in view of the fact that we did not succeed in getting a satisfactory lower bound of $F(n)$ we do not investigate this question at present.

To prove (6) consider again the family of $[n^2/4]$ sets $A_{r,s}$ used in the proof of (2).

We already showed that the largest union-free subfamily of our sets has n elements. Thus

$$g(n) \geq [\frac{n^2}{4}]/n \geq \frac{n}{4} .$$

Hanson suggested that a more careful analysis of this family would in fact give $g(n) \geq n/3$.

Now finally we prove (7). Consider again the sets $A_{r,s}$ used in the proof of (2). As stated before the number of these sets is $[n^2/4]$ and the sets $A_{r,s}$ correspond to a complete bipartite graph of n vertices with $[n/2]$ white and $[(n+1)/2]$ black vertices. If a subfamily $\{A_{r_i,s_i}\}$ is such that no four distinct elements of it satisfy (3), then as stated the corresponding bipartite graph (we join the vertices r_i and s_i) has no rectangle. By a theorem of Reiman [2] such a graph can have at most

$$(1 + o(1))n^{3/2}/2\sqrt{2}$$

edges; thus we immediately obtain

$$G(n) > (1 + o(1))n^{1/2}/2^{1/2} .$$

Now we prove the upper bound of (7). Let q be the smallest prime power for which $q^2 + q + 1 \geq n$. By a well known result of Singer [3], there are $q + 1$ residues $\mod (q^2+q+1)$, $a_1, \ldots, a_{q+1}$, so that all non zero residue classes have a unique representation in the form $a_j - a_i$.

Now we split the subsets of a set $\mathcal{Q}$ of $q^2 + q + 1 \geq n$ elements into $q + 1$ classes so that the sets of none of the classes contain four sets satisfying (3). To see this put in the i-th class $(1 < i < q + 1)$ all sets having $\overline{a_j - a_i}$ $(1 \leq j \leq q + 1, i \neq j)$ elements, where $\overline{a_j - a_i}$ is the least positive integer $\equiv (a_j - a_i) \pmod{q^2+q+1}$.

If four distinct sets A_1, A_2, A_3, A_4 of the i-th class satisfy (3) we would have

$$|A_1| + |A_2| = |A_3| + |A_4|$$

or

$$a_{j_1} - a_i + a_{j_2} - a_i = (a_{j_3} - a_i + a_{j_4} - a_i) \pmod{q^2+q+1}.$$

Hence, $a_{j_1} - a_{j_3} = (a_{j_4} - a_{j_2}) \pmod{q^2+q+1}$, which is impossible.

Thus

$$G(n) \le G(q^2+q+1) \le q + 1 \le \sqrt{n} + 1,$$

which completes the proof of (7).

References

1. P. Erdös and J. Komlós, On a problem of Moser, Combinatorial Theory and its Applications I, (Edited by P. Erdös, A. Rényi, and V. T. Sós), North Holland Publishing Company, Amsterdam, 1969.

2. I. Reiman, Über ein Problem von K. Zarankiewicz, Acta Math. Acad. Sci. Hungarica 9 (1958), 269-273.

3. J. Singer, A theorem in finite projective geometry and some applications to number theory, Trans. Amer. Math. Soc. 43 (1938), 377-385.

THE ROLE OF GRAPH THEORY IN SOME SIEVE
ARGUMENTS OF PROBABILITY THEORY

Janos Galambos
Temple University
Philadelphia, PA 19122

1. Motivation of the problem. Let A_1, A_2, ..., A_n be events in the sense of probability theory and let B_r denote the event that exactly r of the A's occur. A frequent problem is to evaluate the probability $P(B_r)$ of B_r in terms of the probabilities of products of the A's. As an example, consider the following problem. Let Y_j be the amount an insurance company has to pay out on the j-th day of the year. Evidently, Y_j varies from day to day and the company should be able to pay the largest claim, but it is a loss in financial terms if a much larger amount is kept in the bank account of the company than needed for the operations. That is, the company is interested in the behaviour of $Z_n = \max(Y_1, ..., Y_n)$. Let $A_j = \{Y_j \geq x\}$, where x is a given number. Then $B_0 = \{$none of the A_j's occurs$\} = \{Y_j < x$ for all $j\} = \{Z_n < x\}$. A similar situation arises if Y_j denotes the time needed for servicing the j-th component of a machine, where $j = 1, 2, ..., n$. Again, if $A_j = \{Y_j \geq x\}$, then $B_0 = \{Z_n < x\}$. Evidently, Z_n is the time needed for the completion of the service of the machine, if servicing starts on all components at the same time. In both cases, B_1 describes the behaviour of the second largest among the Y's, and in general, B_r is related to the $(r + 1)$-st largest of the Y's. Usually values of

$$P(A_{i_1} A_{i_2} \cdots A_{i_k}) \tag{1}$$

can be estimated for all $1 \leq i_1 < i_2 < \cdots < i_k \leq n$. Therefore we seek formulas for $P(B_r)$ in terms of the expressions in (1). The following is a well-known exact formula due to K. Jordán (see Takács [7]):

$$P(B_r) = \sum_{k=0}^{n-r} (-1)^k \binom{k + r}{r} \sum P(A_{i_1} A_{i_2} \cdots A_{i_{k+r}}), \tag{2}$$

where the inner sum is over all $(k + r)$-vectors $(i_1, i_2, \ldots, i_{k+r})$, where $1 \leq i_1 < i_2 < \cdots < i_{k+r} \leq n$. The disadvantage of (2) is that it contains a large number of terms. Thus if the values (1) are available only with an error term, they add up to a larger value than the major term (the signs $(-1)^k$ cannot be taken into account for the errors). Therefore let us try to restrict the number of terms in (2) by dropping several of them; this naturally results in inequalities instead of the equality in (2). In other words, let H_1 and H_2 be two sets of subsets of $\{1, 2, \ldots, n\}$ such that for any $s \geq 0$,

$$(3) \quad \sum_{k=0}^{2s+1} (-1)^k \binom{k + r}{r} \sum_{H_1} P(A_{i_1} A_{i_2} \cdots A_{i_{k+r}})$$

$$\leq P(B_r) \leq \sum_{k=0}^{2s} (-1)^k \binom{k + r}{r} \sum_{H_2} P(A_{i_1} A_{i_2} \cdots A_{i_{k+r}}),$$

where $\sum_{H_j}$ denotes summation over all $(k + r)$-vectors $(i_1, i_2, \ldots, i_{k+r}) \in H_j$. A general result in probability theory – the method of indicators of Loeve [5] – yields that (3) holds, if it holds, whenever the A's are replaced by the "sure" event and the "impossible" event. These choices make all probabilities zero or one; hence (3) reduces to counting the terms of the two sides of (3). Rényi [6] introduced an arbitrary graph G with vertices $\{1, 2, \ldots, n\}$ and let H_1 and H_2 be sets of vertices of certain types of subgraphs of G. The above mentioned fact therefore reduces (3) to counting the number of certain subgraphs of a given graph. Rényi [6] gave criteria for H_1 and H_2 for (3) to hold when $r = 0$; and for arbitrary r, a general result was obtained in Galambos [1]. These results helped the author to obtain new stochastic models for the situations mentioned earlier in this section (see [2], [3] and [4]).

In the present paper inequalities related to (3) will be discussed without further reference to probability theory. The probabilistic applications of the improvements on the inequalities of Galambos [1] will be presented elsewhere. The author would like to point out that slight improvements on these inequalities yield essential improvements in their applicability.

2. The graph theoretical inequalities related to (3). Consider a graph $G = (H, E)$ with vertices $H = \{1, 2, \ldots, n\}$ and edge set E. For a set $h \subset H$, the graph $G(h) = (h, E(h))$ is the subgraph of G induced by h. Let A be a finite set and let N_A denote the number of elements of A. In particular, N_H is the number of vertices and N_E is the number of edges of G. For a given graph $G = (H, E)$, if $A = \{h: h \subset H, N_h = k, N_{E(h)} \leq j\}$, we then put $N_G(j,k) = N_A$.

The aim of the present paper is to prove the following result.

THEOREM. Let Γ be a collection of graphs such that if $G = (H, E)$ belongs to Γ, so does $G(h)$ for all $h \subset H$. Let $T(r) \geq 1$, $r = 0, 1, 2, \ldots$ be a non-decreasing sequence of positive integers. Assume that if $G \in \Gamma$ and if $N_H = r$ then

$$N_G(T(r), r) = 1. \tag{4}$$

Let us further assume that for any $G = (H, E)$ belonging to Γ, if $h \subset H$ with $N_h = r$ and if $E(h)$ is empty, then for any $h_1 \supset h$ with $N_{h_1} = r + 1$, $E(h_1) \leq T(r)$. Then for any $G \in \Gamma$ and for $r = 0, 1, 2, \ldots$ and $s \geq 0$,

$$\sum_{k=0}^{s+1} \binom{2k+r}{r} N_G(T(r), 2k+r) - \sum_{k=0}^{s} \binom{2k+1+r}{r} N_G(0, 2k+1+r) \geq \delta_{n,r} \tag{5}$$

and

$$\sum_{k=0}^{s} \binom{2k+r}{r} N_G(0, 2k+r) - \sum_{k=0}^{s} \binom{2k+1+r}{r} N_G(T(r), 2k+1+r) \leq \delta_{n,r} \tag{6}$$

where $n = N_H$ and $\delta_{n,r} = 1$ or 0 according as $n = r$ or $n \neq r$.

Before giving the proof, let us make a few remarks. First of all we introduce the convention that the empty set is a set with zero elements; hence for $r = 0$ (4) holds whatever the value of $T(0)$. Note that the additional assumption immediately following (4) is satisfied for all graphs if $T(r) \leq r$; but for smaller values of $T(r)$, it excludes certain graphs. Another restriction is (4), which relates the choices of Γ and the sequence $T(r)$. In particular, if Γ is the set of all finite graphs, then, applying (4) to the complete graph with

r vertices, we have that $T(r) \geq \binom{r}{2}$. It turns out that for this
case equality holds here for $r \geq 3$; more precisely, in [1] the
author proved that $T(r) = \max(1, r, \binom{r}{2})$, if Γ includes all finite
graphs. From the proof there it is seen that $T(r)$ cannot be made
smaller for the class of all graphs; hence in order to improve on the
inequalities of [1] by reducing the value of $T(r)$, Γ must necessarily exclude some graphs.

In order to prove our theorem, we need the following lemmas.

LEMMA 1. Let $G = (H, E)$ be such that each of its vertices is
adjacent to exactly $N_H - r - 1$ other vertices of G. Then for any
choice of $T(r) \geq 0$, inequality (5) holds.

Proof. Note that the left hand side of (5) now contains at most
one negative term, corresponding to $k = 0$. Namely, by the assumption on G, any subgraph with more than $r + 2$ vertices contains
edges. Since $\delta_{N_H,r} = 0$ by assumption, there is nothing to prove if
$N_G(0, r + 1) = 0$. Let us assume that $h_1, h_2, \ldots, h_t$ are subsets of
H with $N_{h_j} = r + 1$, such that $E(h_j) = \phi$, $j = 1, 2, \ldots, t$. By
the definition of h_j and by the construction of G, each element of
h_j is connected to all of those elements of H which are not contained in h_j itself. Therefore the sets $h_1, h_2, \ldots, h_t$ are disjoint and thus their contribution to $N_G(T(r), r)$ is $t(r + 1)$.
Hence if $N_G(0, r + 1) = t$, the left hand side of (5) is non-negative,
independently of the choice of $T(r)$. The lemma is established.

LEMMA 2. Let $G = (H, E)$ be such that if $h \subset H$ with $N_h = r$
for which $E(h)$ is empty, then for any $h_1 \supset h$ with $N_{h_1} = r + 1$,
$E(h_1) \leq T(r)$. Then (6) holds whenever G belongs to one of the following two types of graphs: either $N_H = r + 1$ and G has no isolated points or $N_H > r$ is arbitrary and each vertex is connected to
exactly $N_H - r$ other vertices of G.

Proof. In both cases there is at most one positive term $N_G(0,r)$
on the left hand side of (6). In the first case, with $N_H = r + 1$,
we have in addition that $N_G(0,r) \leq 1$. But by assumption $N_E \leq T(r)$;
thus $N_G(T(r), r + 1) = 1$, and (6) is evident. Turning to the second
case, we proceed as in the proof of Lemma 1. Let $N_G(0, r) = t$, and
let $h_1, h_2, \ldots, h_t$ be subsets of H with $N_{h_j} = r$ and for which

$E(h_j)$ are empty. As we have seen, h_1, h_2, ..., h_t are disjoint. For h_j take an additional element of H and consider the corresponding subgraph. By the assumption on its number of edges, this contributes one to $N_G(T(r), r + 1)$. Since the h's are disjoint, all these new subgraphs are distinct for $r > 1$; hence (6) holds again for $r > 1$. For $r = 1$, a direct check is immediate, and the lemma is proved.

We remark that for certain $T(r)$, the family of the graphs of the second type may be empty. Our aim was simply to make it possible in the proof of the theorem to exclude these comparatively simple cases.

We now turn to the proof of the theorem.

Proof of the main theorem. By the results of [1] and [6] we get that (5) and (6) hold for $r = 0$ and $r = 1$ for any class Γ and for any choice of $T(r) \geq 1$. We can therefore use induction over r. Assume that (5) and (6) have been proved for $r - 1$ ($r \geq 2$) for all elements of Γ and let us prove (5) for r. An additional induction will be employed, namely, over N_H. For $N_H < r$, (5) is evident and for $N_H = r$, (4) implies (5). Assume now that (5) is proved for r and for $N_H = n$, and consider an arbitrary element of Γ with $N_H = n + 1$. Let $H = \{1, 2, ..., n+1\}$, and consider the following subgraphs of $G = (H, E)$. Let $H' = \{1, 2, ..., n\}$, let H'' be that subset of elements of H' which are adjacent with the vertex $n + 1$ in G, and finally let H''' be that subset of elements of H' which are not adjacent with the vertex $n + 1$. We put $G' = G(H')$, $G'' = G(H'')$ and $G''' = G(H''')$, and let n'' and n''' denote, respectively, the number of elements of H'' and H'''. By Lemma 1, we can label the vertices so that $\delta_{n''',r} = 0$. Distinguishing subgraphs of G according as they contain $n + 1$ or not, we evidently have

$$N_G(T(r), 2k+r) \geq N_{G'}(T(r), 2k+r) + N_{G'''}(T(r), 2k+r-1)$$

and

$$N_G(0, 2k+1+r) = N_{G'}(0, 2k+1+r) + N_{G'''}(0, 2k+r).$$

Multiplying these equations by $\binom{2k + r}{r}$ and $\binom{2k + 1 + r}{r}$, respectively and summing with respect to k, and then subtracting the last identity from the inequality obtained, we have

$$\sum_{k=o}^{s+1} \binom{2k + r}{r} N_G(T(r), 2k+r) - \sum_{k=o}^{s} \binom{2k + 1 + r}{r} N_G(0, 2k+1+r)$$

$$\geq \sum_{k=o}^{s+1} \binom{2k + r}{r} N_{G'}(T(r), 2k+r) - \sum_{k=o}^{s} \binom{2k + 1 + r}{r} N_{G'}(T(r), 2k + 1 + r)$$

$$+ \sum_{k=o}^{s+1} \binom{2k + r}{r} N_{G'''}(T(r), 2k+r-1) - \sum_{k=o}^{s} \binom{2k + 1 + r}{r} N_{G'''}(0, 2k+r).$$

Applying the elementary identity $\binom{m + 1}{k + 1} = \binom{m}{k + 1} + \binom{m}{k}$ to the last two sums and then the induction hypothesis, first over r and then over n, we obtain

$$\sum_{k=o}^{s+1} \binom{2k + r}{r} N_G(T(r), 2k+r) - \sum_{k=o}^{s} \binom{2k + 1 + r}{r} N_G(0, 2k+1+r)$$

$$\geq \delta_{n',r} - \delta_{n''',r} + \delta_{n''',r-1},$$

which, by the choice of $n + 1$, and thus of G''', yields (5). The proof of (6) is similar, only the vertex $n + 1$ should be selected in view of Lemma 2. The proof is thus complete.

References

1. J. Galambos, On the sieve methods in probability theory I, _Studia Sci. Math. Hungar._ 1 (1966), 39-50.

2. _____, On the sieve methods in probability theory II, _Ghana J. Sci._ 10 (1970), 11-15.

3. _____, On the distribution of the maximum of random variables, _Ann. Math. Statist._ 43 (1972), 516-521.

4. _____, The distribution of the maximum of a random number of random variables with applications, _J. Appl. Prob._ 10 (1973), (to appear in the March issue).

5. M. Loeve, Sur les systemes d'evenements, _Ann. Univ. Lyon_, Sect. A. 5 (1942), 55-74.

6. A. Rényi, A general method to prove theorems of probability theory and some of its applications (in Hungarian), <u>Magyar</u> <u>Tud</u>. <u>Akad</u>. <u>Mat</u>. <u>Fiz</u>. <u>Oszt</u>. <u>Közl</u>. 11 (1961), 79-105.

7. L. Takács, On the method of inclusion and exclusion, <u>J</u>. <u>Amer</u>. <u>Statist</u>. <u>Assoc</u>. 62 (1967), 102-113.

THE PFAFFIAN AND 1-FACTORS OF GRAPHS II

P. M. Gibson
University of Alabama
Huntsville, AL 35807

One method of enumerating the 1-factors of a graph G with an even number of points is to affix $\pm$ signs to the elements of the adjacency matrix of G so that a skew-symmetric matrix is obtained for which the pfaffian is equal to the number of 1-factors of G. (For a definition of the pfaffian of a matrix see [6].) This technique has been successful for a number of graphs that arise in the dimer problem of statistical mechanics. Pla [5] (also see [4]) characterizes graphs for which the 1-factors can be enumerated in this way. The author [2] has considered the relationship between the number of lines and the number of points of such graphs. In this paper, we continue our study of this relationship.

We adopt the basic graph terminology of Harary [3]. The number of 1-factors and the number of lines of a graph G are denoted by $\mu(G)$ and $\varepsilon(G)$, respectively, and the pfaffian of an even order skew-symmetric matrix B by $\varphi(B)$. If G is a graph with an even number of points and $\pm$ signs can be affixed to the elements of the adjacency matrix A of G so that a skew-symmetric matrix B is obtained such that $\mu(G) = \varphi(B)$, then we call G a _pfaffian graph_. For each integer $n \geq 2$, let Γ_n be the set of all pfaffian graphs with $2n$ points, and let

$$\alpha_n = \max\{\varepsilon(G) \mid G \in \Gamma_n\},$$

$$\beta_n = \max\{\varepsilon(G) \mid G \in \Gamma_n, \mu(G) > 0\}.$$

It is not difficult to show that $\alpha_n = n(n+1)$ or $\alpha_n = (n-1)(2n-1)$ according to whether $2 \leq n \leq 3$ or $n > 3$. The determination of β_n is more interesting. The author [2] has proposed the following.

CONJECTURE 1. _For each integer_ $n \geq 2$, $\beta_n = n(n+1)$.

It has been shown [2] that this conjecture is true for each $2 \leq n \leq 6$, and that $n(n + 1) \leq \beta_n \leq (9n^2 + 2n + 1)/8$ for each $n > 6$. In this paper, we prove that Conjecture 1 is true for each $2 \leq n \leq 7$, and that $n(n + 1) \leq \beta_n \leq (11n^2 + 3n + 2)/10$ for each $n > 7$.

Using results on conversion of the permanent into the determinant [1], we have solved the analogous problem for bipartite graphs [2]. For each positive integer n, let F_n be a bipartite graph with $2n$ points $\{u_1, \ldots, u_n, v_1, \ldots, v_n\}$ such that u_i and v_j are adjacent if and only if $i + j \geq n$.

THEOREM 1 [2]. <u>Let</u> G <u>be a bipartite pfaffian graph with</u> $2n$ <u>points. If</u> $\mu(G) > 0$, <u>then</u>

$$\varepsilon(G) \leq (n^2 + 3n - 2)/2,$$

<u>with equality if and only if</u> G <u>is isomorphic to</u> F_n.

From Theorem 1 and [2, Lemma 1], we obtain the following:

LEMMA 1. <u>Let</u> G <u>be a bipartite pfaffian graph which is isomorphic to a spanning subgraph of the complete bipartite graph</u> $K_{n,n}$, <u>where</u> $n \leq 5$. <u>If</u> G <u>has no isolated points, then</u>

$$\varepsilon(G) \leq (n^2 + 3n - 2)/2,$$

<u>with equality if and only if</u> G <u>is isomorphic to</u> F_n.

If k and n are positive integers with $k < n$, let H_n^k be a graph with $2n$ points $\{v_1, \ldots, v_{2n}\}$ such that if $i \neq j$, then v_i and v_j are adjacent if and only if $i + j \geq 2n$ or $i, j \in [n - k, n]$.

LEMMA 2. <u>For each integer</u> $n \geq 3$, H_n^1 <u>and</u> H_n^2 <u>are pfaffian graphs.</u>

<u>Proof.</u> Since it is shown in the proof of Theorem 2 of [2] that H_n^1 is a pfaffian graph, we only consider H_n^2. Let $A_n = [a_{ij}]$ be the adjacency matrix of H_n^2, and let $B_n = [b_{ij}]$ be a $(2n)$-square skew-symmetric matrix such that for all $i < j$,

$$
b_{ij} = \begin{cases} -a_{ij} & \text{if } i + j = 2n \text{ or } \cdot i,j \in \{n + 1, n + 2\}, \\[2em] a_{ij} & \text{otherwise.} \end{cases}
$$

It is easy to show by induction that

$$
\mu(H_n^2) = 2^{n-3}7 = \varphi(B_n), \quad n = 3, 4, \ldots .
$$

Let G be a graph with an even number of points $\{v_1, \ldots, v_{2n}\}$. Denote the degree of v_i by δ_i, let G_{ij} be the graph that remains after v_i and v_j (and all lines incident with these points) are removed, and let $\sigma_{ij} = \delta_i + \delta_j - 1$. If $\{i_1, \ldots, i_n, j_1, \ldots, j_n\} = \{1, \ldots, 2n\}$, let $G\{i_1, \ldots, i_n\}$ be the maximal spanning bipartite subgraph of G joining $\{v_{i_1}, \ldots, v_{i_n}\}$ and $\{v_{j_1}, \ldots, v_{j_n}\}$. Three lemmas will now be developed, and the last one of these will be used to prove Conjecture 1 for $n = 7$.

LEMMA 3. Let G be a pfaffian graph with 8 points, none of which is isolated. Then

$$
\varepsilon(G) \le 20, \tag{1}
$$

with equality only if G is isomorphic to H_4^1 or H_4^2.

Proof. The inequality follows from [2, Lemma 4]. Suppose that equality holds in (1). Let $A = [a_{ij}]$ be the adjacency matrix of G. If $a_{ij} = 1$, then G_{ij} is a pfaffian graph [2, Lemma 3], $\varepsilon(G_{ij}) \le 12$ [2, Lemma 4], and

$$
20 = \varepsilon(G) = \sigma_{ij} + \varepsilon(G_{ij}) \le \sigma_{ij} + 12.
$$

Hence,

$$
a_{ij} = 1 \Rightarrow \sigma_{ij} \ge 8, \quad i,j = 1, \ldots, 8. \tag{2}
$$

Therefore, since G has no isolated point, $\delta_i \ge 2$ for $i = 1, \ldots, 8$. Since every graph isomorphic to a pfaffian graph is a pfaffian graph, we may assume that

$$
2 \le \delta_1 \le \delta_2 \le \ldots \le \delta_8. \tag{3}
$$

Suppose that $\delta_1 = 2$. Then, since (2) and (3) hold, $a_{1j} = 1$ and $\sigma_{1j} = 8$ for $j = 7,8$. Hence, G_{18} is a pfaffian graph with $\ell(G_{18}) = 12$. It is not difficult to show that each pfaffian graph with 6 points and 12 lines is either a regular graph of degree 4 or is isomorphic to H_3^1 or H_3^2. Therefore, since G_{18} has a point of degree 5, it is isomorphic to H_3^1 or H_3^2. Hence, G is isomorphic to H_4^1 or H_4^2, if $\delta_1 = 2$. Suppose that $\delta_1 > 2$. We shall show that this leads to a contradiction. Suppose that $\delta_1 = \delta_2 = 3$, $\delta_3 = 4$, and $\delta_i = 6$ for $i = 4, \ldots, 8$. From (2), we have $a_{ij} = 0$ for $i,j = 1, 2, 3$. If we choose $4 \le k \le 8$ such that $a_{2k} = 0$, then $G\{1, 2, 3, k\}$ is isomorphic to a bipartite graph J with adjacency matrix $C = [c_{ij}]$, where

$$C = \begin{bmatrix} 0 & C_o \\ C_o^T & 0 \end{bmatrix}, \qquad C_o = \begin{bmatrix} 1 & 1 & 0 & 0 \\ 0 & 1 & 1 & 1 \\ 1 & 0 & 1 & 1 \\ 1 & 1 & 1 & 1 \end{bmatrix}.$$

Since each spanning subgraph of a pfaffian graph is a pfaffian graph [2, Lemma 2], J is a pfaffian graph. Hence, there exists an 8-square skew-symmetric matrix

$$B = [b_{ij}] = \begin{bmatrix} 0 & B_o \\ -B_o^T & 0 \end{bmatrix}$$

such that

$$b_{ij} = \pm c_{ij}, \quad i,j = 1, \ldots, 8,$$

$$\mu(J) = \varphi(B) = \det B_o,$$

where $\det B_o$ denotes the determinant of B_o. This is a contradiction, since it is not difficult to show that

$$\det B_o \le 6 < 8 = \mu(J).$$

We consider 11 other cases. In each of these, we point out a spanning bipartite subgraph L of G which is isomorphic to a spanning subgraph of $K_{4,4}$ and satisfies the following properties:

(a)   L has no isolated point,

(b) $\varepsilon(L) \geq 13$,

(c)   L is not isomorphic to F_4.

The existence of such a pfaffian graph L contradicts Lemma 1. In each case, we omit the details of showing that properties (a), (b), and (c) are satisfied.

<u>Case</u> 1. $\delta_3 + \delta_4 \leq 8$. Using (2) and (3), we see that $a_{ij} = 0$ for $i,j = 1, \ldots, 4$, and we let $L = G\{1,\ldots,4\}$.

<u>Case</u> 2. $\delta_3 = 3$, $\delta_4 = 6$. Since $\delta_1 \geq 3$, (3) implies that $\delta_8 = 7$, and $\delta_i = 6$ for $i = 5, 6, 7$. Let $L = G\{1,\ldots,4\}$.

<u>Case</u> 3. $\delta_2 = 3$, $4 \leq \delta_3 \leq 5$, $\delta_4 = 5$. From (2) and (3), we have $a_{1j} = a_{2j} = 0$ for $j = 1, \ldots, 4$, and we let $L = G\{1,\ldots,4\}$.

<u>Case</u> 4. $\delta_1 = 3$, $\delta_2 = \delta_3 = 4$, $\delta_4 = \delta_5 = 5$, $\delta_6 = 6$. Then $a_{23} = 0 = a_{1j}$ for $j = 2, \ldots, 5$. If $a_{ik} = 0$ for some $2 \leq i \leq 3 < k \leq 5$, let $L = G\{1,2,3,k\}$. If $a_{24} = a_{25} = a_{34} = a_{35} = 1$, let $L = G\{1,4,5,8\}$.

<u>Case</u> 5. $\delta_1 = 3$, $\delta_2 = \delta_3 = 4$, $\delta_4 = 5$, $\delta_5 = 6$. Then $a_{14} = 0 = a_{ij}$ for $i,j = 1, 2, 3$, and $a_{1k} = 0$ for some $5 \leq k \leq 8$. If $a_{m4} = 0$ for some $2 \leq m \leq 3$, let $L = G\{1, m, 4, k\}$. If $a_{24} = 1 = a_{34}$, then $a_{4p} = 0$ for some $6 \leq p \leq 8$, and we let $L = G\{1,4,k,p\}$.

<u>Case</u> 6. $\delta_1 = 3$, $\delta_2 = 4$, $\delta_3 = 5 = \delta_5$, $\delta_6 = 6$. Then $a_{1j} = 0$ for $j = 2, \ldots, 5$. If $a_{ik} = 0$ for some $3 \leq i < k \leq 5$, then let $L = G\{1,3,4,5\}$. If $a_{ik} = 1$ for all $3 \leq i < k \leq 5$, then $a_{mp} = 0$ for some $3 \leq m \leq 5 < p \leq 8$, and we let $L = G\{m,6,7,8\}$.

<u>Case</u> 7. $\delta_1 = 4 = \delta_3$, $\delta_4 = 5$. Then $a_{ij} = 0$ for $i,j = 1, 2, 3$, and $a_{1k} = 0$ for some $4 \leq k \leq 8$. Let $L = G\{1,2,3,k\}$.

<u>Case</u> 8. $\delta_1 = 4 = \delta_2$, $\delta_3 = 5 = \delta_7$. Then $a_{12} = 0$ and $a_{1i} = 0 = a_{1j}$ for some $3 \leq i < j \leq 7$. Let $L = G\{1,i,j,8\}$.

<u>Case</u> 9. $\delta_1 = 4 = \delta_2$, $\delta_3 = 5 = \delta_6$, $\delta_7 = 6$. Then $a_{12} = 0$. If there exist distinct $i,j,k \in \{3,\ldots,8\}$ such that $a_{ij} = 0 = a_{jk}$, then there exists $m \in \{3,\ldots,8\} - \{i,j,k\}$ such that we can let $L = G\{i,j,k,m\}$. If there do not exist distinct $i,j,k \in \{3,\ldots,8\}$ such that $a_{ij} = 0 = a_{jk}$, then there exist $3 \leq m < p \leq 6$ such that $\varepsilon(G\{m,p,7,8\}) \geq 14$, and we let $L = G\{m,p,7,8\}$.

<u>Case</u> 10. $\delta_1 = 4$, $\delta_2 = 5$. There exist distinct $i,j,k \in \{2,\ldots,7\}$ such that $a_{ij} = 0 = a_{jk}$. Let $L = G\{i,j,k,8\}$.

<u>Case</u> 11. $\delta_1 = 5$. There exist distinct $i,j,k,m \in \{1,\ldots,8\}$ such that $a_{ij} = a_{jk} = a_{km} = 0$. Let $L = G\{i,j,k,m\}$.

LEMMA 4. <u>Let</u> G <u>be a pfaffian graph with</u> 10 <u>points, none of which is isolated. Then</u>

(4) $\qquad\qquad\qquad\qquad \varepsilon(G) \le 30,$

<u>with equality only if one of the following holds</u>:

     (a)   G <u>is isomorphic to</u> H_5^1 <u>or</u> H_5^2,

     (b)   G <u>has two points of degree</u> 1,

     (c)   G <u>has six points of degree</u> 4.

<u>Proof</u>. The inequality follows from [2, Lemma 4]. Suppose that equality holds in (4). Let $A = [a_{ij}]$ be the adjacency matrix of G. If $\delta_i = 1$ for some $1 \le i \le 10$, then it follows from Lemma 3 that G has two points of degree 1. If $\delta_i \ge 2$ for all $1 \le i \le 10$, while $\delta_k = 2$ for some $1 \le k \le 10$, then it follows from Lemma 3 that G is isomorphic to H_5^1 or H_5^2. Suppose that $\delta_i \ge 3$ for $i = 1, \ldots, 10$. If $a_{ij} = 1$, then G_{ij} is a pfaffian graph with no isolated point. Hence, by Lemma 3, if $a_{ij} = 1$, then $30 = \sigma_{ij} + \varepsilon(G_{ij}) \le \sigma_{ij} + 20$. Therefore,

(5) $\qquad\qquad a_{ij} = 1 \Rightarrow \sigma_{ij} \ge 10$, $i,j = 1, \ldots, 10$.

Suppose that $a_{km} = 1$ and $\sigma_{km} = 10$. It follows from Lemma 3 that G_{km} is isomorphic to H_4^1 or H_4^2. We may assume that $\delta_k < \delta_m$, $G_{km} = H_4^1$ or H_4^2, $k = 1$, and $m = 10$. Since $\delta_1 \le 5$ and $\delta_2 \le 4$, (5) implies that $a_{12} = 0$. Hence, $a_{2,10} = 1$ and $\delta_2 = 3$. Therefore, by (5), $\delta_{10} \ge 8$. Hence, $\delta_1 = 3$. Since $\delta_1 = 3$ and $G_{1,10} = H_4^1$ or H_4^2, (5) implies that $a_{1j} = 0$ for $j = 1, \ldots, 5$. Therefore, if we let $J = G\{1, \ldots, 5\}$, then it follows that J is a bipartite pfaffian graph for which $\varepsilon(J) \ge 19$, $\mu(J) > 0$, and J is not isomorphic to F_5. This contradiction to Theorem 1 proves that

$$a_{ij} = 1 \Rightarrow \sigma_{ij} \geq 11, \quad i,j = 1, \ldots, 10. \tag{6}$$

We may assume that

$$3 \leq \delta_1 \leq \delta_2 \leq \cdots \leq \delta_{10}. \tag{7}$$

If $\delta_4 + \delta_5 \leq 11$, then (6) and (7) imply that $a_{ij} = 0$ for $i,j = 1, \ldots, 5$. Hence, if $J = G\{1,\ldots,5\}$, then J is a bipartite pfaffian graph which is isomorphic to a spanning subgraph of $K_{5,5}$, and $\varepsilon(J) \geq 20$. Therefore, by Lemma 1, J must have an isolated point. Hence, $a_{ij} = 0$ for $i,j = 1, \ldots, 6$, and G has six points of degree 4. Now suppose $\delta_4 + \delta_5 \geq 12$. Using (6), it is not difficult to show that there are only two cases to consider.

<u>Case 1.</u> $\delta_1 = 3$, $\delta_4 = 6$, $\delta_8 = 9$. If $J = G\{1,\ldots,5\}$, then it is easy to show that J is a bipartite pfaffian graph with $\varepsilon(J) \geq 19$, $\mu(J) > 0$, and J is not isomorphic to F_5. This contradicts Theorem 1.

<u>Case 2.</u> $\delta_1 = 6$. Since G has 10 points and is regular of degree 6, there exist distinct i,j,k,m, and p such that $a_{ij} = a_{jk} = a_{km} = a_{mi} = a_{mp} = 0$ or $a_{ij} = a_{jk} = a_{km} = a_{mp} = a_{pi} = 0$. If $L = G\{i,j,k,m,p\}$, then L is a bipartite pfaffian graph which is isomorphic to a spanning subgraph of $K_{5,5}$, $\varepsilon(L) \geq 20$, and L has no isolated point. This contradiction to Lemma 1 completes our proof of Lemma 4.

LEMMA 5. <u>Let</u> G <u>be a pfaffian graph with</u> 12 <u>points and</u> $\mu(G) > 0$. <u>Then</u>

$$\varepsilon(G) \leq 42, \tag{8}$$

<u>with equality only if</u> G <u>is isomorphic to</u> H_6^1 <u>or</u> H_6^2.

<u>Proof.</u> The inequality follows from [2, Theorem 2]. Suppose that equality holds in (8). Since $\mu(G) > 0$, Lemma 4 implies that $\delta_i \geq 2$ for $i = 1, \ldots, 12$. Suppose that $\delta_k = 2$. There is some j such that $a_{kj} = 1$ and $\mu(G_{kj}) > 0$. It follows from Lemma 4 that $\delta_j = 11$ and G_{kj} is isomorphic to H_5^1 or H_5^2. Hence, G is isomorphic to H_6^1 or H_6^2. Suppose that $\delta_i \geq 3$ for $i = 1, \ldots, 12$. Using Lemma 4, we obtain

(9) $a_{ij} = 1 \Rightarrow \sigma_{ij} \geq 12$, $i,j = 1, \ldots, 12$.

Suppose that $a_{km} = 1$ and $\sigma_{km} = 12$. Then $\varepsilon(G_{km}) = 30$. Since each point of G has degree at least 3, (9) implies that each point of G_{km} has degree at least 2. It is not difficult to show that since $\mu(G) > 0$, G_{km} cannot have six points of degree 4. Therefore, by Lemma 4, G_{km} is isomorphic to H_5^1 of H_5^2. As in the proof of Lemma 4, we can show that this leads to a contradiction of Theorem 1. Hence,

(10) $a_{ij} = 1 \Rightarrow \sigma_{ij} \geq 13$, $i,j = 1, \ldots, 12$.

We may assume that

(11) $3 \leq \delta_1 \leq \delta_2 \leq \cdots \leq \delta_{12}$.

Since $\mu(G) > 0$, (10), (11), and Theorem 1 imply that $\delta_5 + \delta_2 \geq 14$. From (10) and (11), we see that $5 \neq \delta_1 \neq 6$. There are three cases to consider.

Case 1. $\delta_1 = 3$. If $J = G\{1,\ldots,6\}$, it follows from (10) and (11) that $\varepsilon(J) \geq 27$ and $\mu(J) > 0$. This contradicts Theorem 1.

Case 2. $\delta_1 = 4$. From (10) and (11), we have $a_{ij} = 0$ for $i = 1, \ldots, 4$, and $j = 1, \ldots, 8$. Hence $\varepsilon(G\{1,\ldots,6\}) = 28$ and $\mu(G\{1,\ldots,6\}) > 0$. This contradicts Theorem 1.

Case 3. $\delta_1 = 7$. Then $\delta_i = 7$ for $i = 1, \ldots, 12$. Let $a_{km} = 1$. Then G_{km} is a pfaffian graph with 10 points, $\varepsilon(G_{km}) = 29$, and each point of G_{km} has degree at least 5 and at most 7. It is not difficult to prove that no such graph can exist.

THEOREM 2. <u>For each integer</u> $n \geq 2$,

$$\beta_n = n(n + 1) \quad \underline{if} \quad n \leq 7,$$

$$n(n + 1) \leq \beta_n \leq (11n^2 + 3n + 2)/10 \quad \underline{if} \quad n > 7.$$

Proof. From [2, Theorem 2], we have $\beta_n = n(n + 1)$ if $n \leq 6$, and $\beta_n \geq n(n + 1)$ if $n > 6$. Let G be a pfaffian graph with 14 points and $\mu(G) > 0$. According to [2, Lemma 5], $\varepsilon(G) \leq 57$. Let

$A = [a_{ij}]$ be the adjacency matrix of G. Since $\beta_6 = 42$,

$$(a_{ij} = 1, \ \mu(G_{ij}) > 0) \Rightarrow \sigma_{ij} \geq 15, \quad i,j = 1, \ \ldots, \ 14. \tag{12}$$

Let $\{\{v_{i_1}, v_{j_1}\}, \ \ldots, \ \{v_{i_7}, v_{j_7}\}\}$ be a 1-factor of G. We have

$$\sum_{k=1}^{7} \sigma_{i_k j_k} = 2 \, \varepsilon(G) - 7 = 107.$$

Hence, $\sigma_{i_k j_k} \leq 15$ for some $1 \leq k \leq 7$. Combining this with (12), we see that there exist distinct i and j such that $a_{ij} = 1$, $\mu(G_{ij}) > 0$ and $\sigma_{ij} = 15$. It follows from Lemma 5 that G_{ij} is isomorphic to H_6^1 or H_6^2. We may assume that $\delta_i < \delta_j$, $G_{ij} = H_6^1$ or H_6^2, and $j = 14$. Therefore, since $\delta_2 \leq 4$ and $\mu(G_{12}) > 0$, (12) implies that $a_{12} = 0$. Hence, since $a_{2,13} = 1$ and $\mu(G_{2,13}) > 0$, (12) implies that $a_{2,14} = 1$. Therefore, since $\mu(G_{2,14}) > 0$ and $\delta_2 = 3$, $a_{k,14} = 1$ for $k = 1, \ \ldots, \ 13$, and $G_{2,14}$ is isomorphic to H_6^1 or H_6^2. We have $G_{1,14} = H_6^1$ or H_6^2, while $G_{2,14}$ has two points of degree 11. Hence, $a_{1m} = a_{1p} = 1$ for some $11 \leq m < p \leq 13$. Therefore, $\varepsilon(G\{1,\ldots,7\}) = 35$ and $\mu(G\{1,\ldots,7\}) > 0$. This contradicts Theorem 1. Therefore, $\beta_7 = 56$. It can be shown that $\beta_n \leq (11n^2 + 3n + 2)/10$ for $n > 7$ by a slight modification of the proof of Lemma 5 of [2] (also see [2, Remark 2]).

We propose the following.

CONJECTURE 2. **For each integer** $n \geq 4$, **if** G **is a pfaffian graph with** $2n$ **points and** $\mu(G) > 0$, **then**

$$\varepsilon(G) \leq n(n + 1),$$

with equality if and only if G **is isomorphic to** H_n^1 **or** H_n^2.

It follows from our results that this conjecture is true for $4 \leq n \leq 6$.

References

1. P. M. Gibson, Conversion of the permanent into the determinant, <u>Proc</u>. <u>Amer</u>. <u>Math</u>. <u>Soc</u>. 27 (1971), 471-476.

2. P. M. Gibson, The pfaffian and 1-factors of graphs, <u>Trans</u>. <u>N.Y.</u> <u>Acad</u>. <u>Sci</u>. 34 (1972), 52-57.

3. F. Harary, <u>Graph</u> <u>Theory</u>, Addison-Wesley, Reading, Mass., 1969.

4. C. H. C. Little et J. M. Pla, Sur l'utilisation d'un pfaffien dans l'étude des couplages parfaits d'un graphe, <u>C</u>. <u>R</u>. <u>Acad</u>. <u>Sci</u>. <u>Paris</u> 274 (1972), 447.

5. J. M. Pla, Sur l'utilisation d'un pfaffien dans l'étude des couplages parfaits d'un graphe, <u>C</u>. <u>R</u>. <u>Acad</u>. <u>Sci</u>. <u>Paris</u> 260 (1965), 2967-2970.

6. W. T. Tutte, The factorization of linear graphs, <u>J</u>. <u>London</u> <u>Math</u>. <u>Soc</u>. 22 (1947), 107-111.

ON EMBEDDING GRAPHS IN SQUASHED CUBES

R. L. Graham and H. O. Pollak
Bell Laboratories
Murray Hill, NJ 07974

1. Introduction. For the set of three symbols $S = \{0,1,*\}$, define the function d from $S \times S$ to the nonnegative integers $\mathbb{N}$ by

$$
d(s,s') =
\begin{cases}
1 & \text{if } \{s,s'\} = \{0,1\}, \\
0 & \text{otherwise.}
\end{cases}
$$

For $n \in \mathbb{N}$, d can be extended to a mapping of $S^n \times S^n$ to $\mathbb{N}$ by

$$
d((s_1,\ldots,s_n), (s_1',\ldots,s_n')) = \sum_{k=1}^{n} d(s_k,s_k').
$$

We shall refer to $d((s_1,\ldots,s_n), (s_1',\ldots,s_n'))$ as the distance between the two n-tuples $(s_1,\ldots,s_n)$ and $(s_1',\ldots,s_n')$ although, strictly speaking, this is an abuse of terminology since d does not satisfy the triangle inequality.

For a connected graph G, the distance between two vertices v and v' in G, denoted by $d_G(v,v')$, is defined to be the minimum number of edges in any path between v and v'.

The following problem arose recently in connection with a data transmission scheme of J. R. Pierce [4].

Given a connected graph G, find the least integer $N(G)$ for which it is possible to associate, with each vertex v of G, an element $A(v) \in S^{N(G)}$, such that

$$
d_G(v,v') = d(A(v), A(v')) \tag{1}
$$

for all pairs of vertices v and v' in G.

The mapping A will be called an <u>addressing</u> of G; A(v) will be called the <u>address</u> of the vertex v. Of course, it is not <u>a priori</u> clear that addressings exist for all connected graphs G. It will be seen that an addressing of G is equivalent to a distance-preserving embedding of G into the 1-skeleton of an n-dimensional cube in which certain faces have been "squashed".

In the following sections, various bounds on $N(G)$ are established. In addition, $N(G)$ is determined exactly for a number of classes of graphs.

2. <u>Squashed cubes</u>. Let T_n denote the set of 2^n points $\{(\varepsilon_1,\ldots,\varepsilon_n): \varepsilon_k = 0 \text{ or } 1\}$ in E^n. Let Q_n denote the graph which has T_n as its set of vertices and an edge between the vertices $(\varepsilon_1,\ldots,\varepsilon_n)$ and $(\varepsilon_1',\ldots,\varepsilon_n')$ iff they differ in exactly one coordinate. Thus, Q_n is just the 1-skeleton of an n-cube.

For a given n-tuple $\bar{s} = (s_1,\ldots,s_n) \in S^n = \{0,1,*\}^n$, associate with $\bar{s}$ the set s^* of vertices of Q_n which can be obtained by replacing all s_k which are $*$'s by either 0 or 1. Thus, if $\bar{s}$ has r $*$'s then s^* has 2^r elements. If one replaces the vertices of Q_n which belong to s^* by a single vertex labeled $\bar{s}$ and an edge is placed between $\bar{s}$ and $(\varepsilon_1,\ldots,\varepsilon_n)$ if and only if some element of s^* and $(\varepsilon_1,\ldots,\varepsilon_n)$ are adjacent in Q_n, one forms a new graph Q_n'. One may think of Q_n' as the 1-skeleton of an n-cube in which a certain r-dimensional face was "squashed" and the 2^r vertices were identified by a single vertex.

More generally, if $\bar{s}_1, \ldots, \bar{s}_t$ all belong to S^n and $d(\bar{s}_i,\bar{s}_j) \geq 1$ for $i \neq j$, then one may form the graph Q_n^* by identifying each of the sets of vertices s_k^* by the corresponding single vertex $\bar{s}_k$ with edges incident to $\bar{s}_k$ as previously indicated. Q_n^* may be thought of as the 1-skeleton of an n-cube in which t disjoint hyperfaces have been squashed to points.

An addressing A of a graph G using elements of S^n can now be seen to be equivalent to the existence of a squashed n-cube Q_n^* which is isomorphic to G. The vertex v_k of G corresponds to the vertex $A(v_k)$ in Q_n^* so that A is a distance-preserving map of G onto Q_n^*. $N(G)$ is the least n for which this is possible.

For example, an addressing of K_4, the complete graph on 4 vertices, is given in Figure 1. The associated squashed 3-cube is also shown.

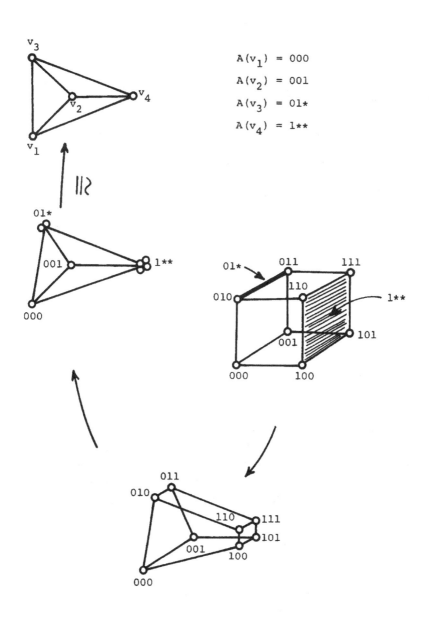

$$A(v_1) = 000$$
$$A(v_2) = 001$$
$$A(v_3) = 01*$$
$$A(v_4) = 1**$$

Figure 1.

3. <u>Upper bounds on</u> $N(G)$. Given an arbitrary finite connected graph G, we first show $N(G) < \infty$. To see this, let

$$N = \sum_{v_i, v_j} d_G(v_i, v_j)$$

where the sum is over all pairs of vertices v_i, v_j of G. Construct the addresses $A(v_k) \in S^N$ as shown below.

$$
\begin{array}{cccc}
 & d_G(v_1, v_2) & d_G(v_1, v_3) & d_G(v_1, v_j) \\
A(v_1) = & \overbrace{0 \ldots\ldots 0} & \overbrace{0 \ldots\ldots 0} & \ldots \quad \overbrace{* \ldots\ldots *} \\
A(v_2) = & 1 \ldots\ldots 1 & * \ldots\ldots * & \ldots \quad * \ldots\ldots * \\
A(v_3) = & * \ldots\ldots * & 1 \ldots\ldots 1 & \ldots \quad * \ldots\ldots * \\
 & \cdot & \cdot & \cdot \\
 & \cdot & \cdot & \cdot \\
A(v_i) = & * \ldots\ldots * & * \ldots\ldots * & \ldots \quad 0 \ldots\ldots 0 \\
 & \cdot & \cdot & \cdot \\
 & \cdot & \cdot & \cdot \\
A(v_j) = & * \ldots\ldots * & * \ldots\ldots * & \ldots \quad 1 \ldots\ldots 1 \\
 & \cdot & \cdot & \cdot \\
 & \cdot & \cdot & \cdot \\
\end{array}
$$

For each pair of vertices v_i, v_j, a unique block of $d_G(v_i, v_j)$ coordinate positions is used to achieve $d(A(v_i), A(v_j)) = d_G(v_i, v_j)$ by placing a block of 0's and a block of 1's in these coordinate positions in v_i and v_j and blocks of *'s in these coordinate positions for all other v_k. This argument shows

(2)
$$N(G) \leq \sum_{v_i, v_j} d_G(v_i, v_j) .$$

If G has n vertices and m_G denotes $\max d_G(v_i, v_j)$, where the max is taken over all v_i, v_j in G, then a slight modification of the preceding argument can be used to show

$$N(G) \le m_G(n-1). \tag{3}$$

No example of a graph G is currently known for which $N(G) > n - 1$. However, it has not yet even been shown that there exists a fixed constant c such that $N(G) \le cn$ for all connected graphs G with n vertices.

4. Lower bounds on $N(G)$. Let $D(G)$ denote the $n \times n$ matrix $(d_{i,j})$, where $d_{i,j} = d_G(v_i, v_j)$, $1 \le i, j \le n$. Let A be an addressing of G using elements of S^N, and write

$$A(v_k) = (s_{k,1}, \ldots, s_{k,N}), \quad 1 \le k \le n.$$

We consider the contributions to the various interpoint distances made by a fixed coordinate position in the addresses, say, the mth coordinate position. An important fact to note is that if $s_{i,m} = 0$ and $s_{j,m} = 1$ then these components contribute 1 to the distance $d(A(v_i), A(v_j))$. Of course, if $s_{i,m} = 1$ and $s_{j,m} = 0$ then these components also contribute 1 to $d(A(v_i), A(v_j))$. In all other cases, these components contribute zero to $d(A(v_i, A(v_j))$. Thus, if $C_m(s)$, $s \in S$, denotes the set of t for which $s_{t,m} = s$ then the mth coordinate position contributes 1 to $d(A(v_i), A(v_j))$ iff either $i \in C_m(0)$, $j \in C_m(1)$ or $j \in C_m(0)$, $i \in C_m(1)$.

This remark can be restated in the following terms. Let $Q(G)$ denote the quadratic form defined by

$$Q(G) = \sum_{1 \le i, j \le n} d_{i,j} x_i x_j ,$$

and let

$$Q'(G) = \sum_{1 \le i < j \le n} d_{i,j} x_i x_j = (1/2) Q(G).$$

Then

$$(4) \qquad Q'(G) = \sum_{m=1}^{N} \Big(\sum_{i \in C_m(0)} x_i \Big) \Big(\sum_{j \in C_m(1)} x_j \Big) .$$

Hence, the existence of an addressing for G using elements of S^N is equivalent to a decomposition of $Q'(G)$ of the type given by (4). A simple algebraic transformation allows (4) to be rewritten as

$$Q'(G) = \frac{1}{4} \sum_{m=1}^{N} \Big[\Big(\sum_{i \in C_m(0)} x_i + \sum_{j \in C_m(1)} x_j \Big)^2$$

(4')

$$- \Big(\sum_{i \in C_m(0)} x_i - \sum_{j \in C_m(1)} x_j \Big)^2 \Big] .$$

This shows that $Q'(G)$ is congruent to a form which has at most N positive squares and at most N negative squares. However, results from matrix theory [3] allow us to conclude from this that

$$N \geq \text{index } Q'(G) = n_+ = \text{number of positive eigenvalues of } D(G)$$

and

$$N \geq \text{index } Q'(G) - \text{rank } Q'(G) = n_- = \text{number of negative eigenvalues of } D(G).$$

We summarize this in the following theorem.[*]

THEOREM. A lower bound for $N(G)$ is given by:

$$(5) \qquad N(G) \geq \max\{n_+, n_-\}.$$

Since the sum of the eigenvalues of $D(G)$ equals the trace of $D(G)$, which is 0, then $\max\{n_+, n_-\} \leq n-1$. Hence, this theorem can never be used to find a counterexample to the inequality $N(G) \leq n-1$.

---

[*] First established in a somewhat different way by H.S. Witsenhausen.

It should be noted that if $\{s_1, \ldots, s_{2^n+1}\} \subseteq S^n$, then for some $i \neq j$, $d(s_i, s_j) = 0$. This implies the bound

$$N(G) \geq \log_2 n \qquad (6)$$

for a graph G with n vertices.

5. Some special cases. We shall apply (5) to determine $N(G)$ for a number of classes of graphs.

(i). $G = K_n$ - the complete graph on n vertices. In this case $d_{i,j} = 1$ for all $i \neq j$ and $n_+ = 1$, $n_- = n-1$. By (5) this implies $N(G) \geq n-1$. However, it is easy to see that $N(G) \leq n-1$ by considering the decomposition of $Q'(G)$ given by

$$Q'(G) = \sum_{1 \leq i < j \leq n} x_i x_j = \sum_{i=1}^{n-1} \left(x_i \sum_{j=i+1}^{n} x_j \right).$$

(This decomposition corresponds to squashing one hyperface of each dimension in the $(n-1)$-cube.) The two inequalities imply

$$N(K_n) = n - 1. \qquad (7)$$

Equation (7) has an interesting graph-theoretic interpretation obtained by associating complete bipartite subgraphs of G with terms of the decomposition of $Q'(G)$ given in (4).

COROLLARY. If K_n is decomposed into t edge-disjoint complete bipartite subgraphs, then $t \geq n - 1$.

No proof of this fact is known which does not use an eigenvalue argument.

(ii). $G = T_n$ - a tree on n vertices. Suppose the vertices of T_n are labeled $v_1, v_2, \ldots, v_n$ so that for $1 \leq k \leq n$, the subgraph of T_n determined by the vertices $v_1, \ldots, v_k$ is a subtree of T_n. In [2], it is shown that it is possible to transform the matrix $D(T_n)$ using elementary row and column operations to a matrix of the form

$$D^*(T_n) = \begin{bmatrix} 0 & 1 & 1 & 1 & \cdots\cdots & 1 & 1 \\ 1 & -2 & 0 & 0 & \cdots\cdots & 0 & 0 \\ 1 & 0 & -2 & 0 & \cdots\cdots & 0 & 0 \\ 1 & 0 & 0 & -2 & \cdots\cdots & 0 & 0 \\ \cdot & \cdot & \cdot & \cdot & & \cdot & \cdot \\ \cdot & \cdot & \cdot & \cdot & & \cdot & \cdot \\ \cdot & \cdot & \cdot & \cdot & & \cdot & \cdot \\ 1 & 0 & 0 & 0 & \cdots\cdots & -2 & 0 \\ 1 & 0 & 0 & 0 & \cdots\cdots & 0 & -2 \end{bmatrix} .$$

Since $D^*(T_n)$ and $D(T_n)$ have the same determinant, then this implies

(8) $$\det D(T_n) = (-1)^{n-1}(n-1)2^{n-2}, \quad n \geq 1,$$

independent of the structure of T_n.

By the way T_n was labeled, the upper left k^{th} order principal submatrices $D_k(T_n)$ of $D(T_n)$ are also distance matrices of trees and, hence, $\det D_k(T_n) = (-1)^{k-1}(k-1)2^{k-2}$, $1 \leq k \leq n$. However, a theorem from linear algebra [3] asserts that, in this case, the number of permanences in sign* of the sequence

$$1, \ \det D_1(T_n), \ \det D_2(T_n), \ \ldots, \ \det D_n(T_n)$$

is just equal to n_+, the number of positive eigenvalues of $D(T_n)$. But there is just one permanence in sign in the above sequence, so that $n_+ = 1$. Since, for $n > 1$, $\det D(T_n) \neq 0$ then $D(T_n)$ is non-singular and $n_- = n-1$. Therefore by (5), $N(T_n) \geq n-1$.

It is easy to show that $N(T_n) \leq n-1$ by inductively addressing the v_k with increasing k, using the fact that if v_i is an exterior vertex of a tree T (i.e., v_i has degree 1) and v_i is adjacent to v_j in T then $d_T(v_i,v_k) = 1 + d(v_j,v_k)$ for all $k \neq i$.

Thus, it follows that

(9) $$N(T_n) = n - 1.$$

In fact, any addressing of T_n with elements of S^{n-1} can use no d's.

---

* Where the sign of 0 may be fixed arbitrarily.

(iii). $G = C_n$ - a cycle on n vertices. Again, (5) can be used directly to show

$$N(C_n) \geq \begin{cases} n - 1 & \text{if } n \text{ is odd,} \\[2mm] \dfrac{n}{2} & \text{if } n \text{ is even.} \end{cases}$$

It is not difficult to construct addressings (cf. [2]) which achieve these bounds so that we have

$$N(C_n) = \begin{cases} n - 1 & \text{if } n \text{ is odd} \\[2mm] \dfrac{n}{2} & \text{if } n \text{ is even.} \end{cases} \tag{10}$$

(iv). $G = Q_n$ - the 1-skeleton of an n-cube. The labeling of Q_n described previously produces an addressing of Q_n using n-tuples of 0's and 1's. On the other hand, by (6), $N(Q_n) \geq \log_2 |Q_n| = n$. This implies $N(Q_n) = n$ (which is not surprising).

(v). $G = K_{n_1, n_2}$ - the complete bipartite graph on n_1 and n_2 vertices. Rearrange the rows and columns of $D(K_{n_1, n_2})$ so that it has the form

$$
D(K_{n_1, n_2}) =
\left[
\begin{array}{cccc|cccc}
0 & 2 & \cdots & 2 & 1 & 1 & \cdots & 1 \\
2 & 0 & \cdots & 2 & 1 & 1 & \cdots & 1 \\
\cdot & & & \cdot & \cdot & \cdot & & \cdot \\
\cdot & & & \cdot & \cdot & \cdot & & \cdot \\
\cdot & & & \cdot & \cdot & \cdot & & \cdot \\
2 & 2 & \cdots & 0 & 1 & 1 & \cdots & 1 \\
\hline
1 & 1 & \cdots & 1 & 0 & 2 & \cdots & 2 \\
1 & 1 & \cdots & 1 & 2 & 0 & \cdots & 2 \\
\cdot & \cdot & & \cdot & \cdot & & & \cdot \\
\cdot & \cdot & & \cdot & \cdot & & & \cdot \\
\cdot & \cdot & & \cdot & \cdot & & & \cdot \\
1 & 1 & \cdots & 1 & 2 & 2 & \cdots & 2
\end{array}
\right].
$$

A straightforward induction argument shows that

(11) $\det D(K_{n_1,n_2}) = (-1)^{n_1+n_2} 2^{n_1+n_2-2} (3n_1 n_2 - 4n_1 - 4n_2 + 4)$

for $n_1, n_2 \geq 1$. By the result mentioned in (ii) on the signs of the determinants of the principal submatrices of $D(G)$, and the fact

$$\det D(K_n) = (-1)^{n-1}(n-1)2^n,$$

it follows that for $D(K_{n_1,n_2})$,

$$n_- = \begin{cases} n_1 + n_2 - 2 & \text{if } n_1 \geq 2, n_2 \geq 2, \\ \\ n_1 + n_2 - 1 & \text{otherwise.} \end{cases}$$

Thus*, by (5),

$$N(K_{n_1,n_2}) \geq \begin{cases} n_1 + n_2 - 2 & \text{if } n_1 \geq 2, n_2 \geq 2, \\ \\ n_1 + n_2 - 1 & \text{otherwise.} \end{cases}$$

In the other direction, the following general result applies.

THEOREM. <u>Suppose</u> G <u>is a graph on</u> n <u>vertices such that for some edge</u> $\{v_i, v_j\}$,

(12) $\min\{d_G(v_i, v_k),\ d_G(v_j, v_k)\} \leq 1$

<u>for all vertices</u> v_k <u>of</u> G. <u>Then</u> $N(G) \leq n - 1$.

Proof. Assume without loss of generality that (12) holds for $i = 1$, $j = 2$. The quadratic form $Q'(G)$ has the following decomposition:

---

[*] $K_{2,2}$ is the only K_{n_1,n_2} for which $D(K_{n_1,n_2})$ is singular.

$$Q'(G) = \sum_{1 \le i < j \le n} d_{ij} x_i x_j$$

$$= (x_1 + \sum_{d_{2i}=2} x_i)(x_2 + \sum_{d_{1j}=2} x_j) + x_3(\ldots) + x_4(\ldots) +$$

$$\ldots + x_n(\ldots),$$

where it is not difficult to check that the appropriate choices can be made in the parenthetical expressions.

Since the complete bipartite graph K_{n_1,n_2} satisfies the hypothesis of the theorem then

$$N(K_{n_1,n_2}) \le n_1 + n_2 - 1.$$

This bound for $N(K_{n_1,n_2})$ has also been obtained in [1]. However, W. T. Trotter has recently shown [5] that

$$N(K_{3n_1+2,3n_2+2}) \le 3n_1 + 3n_2 - 2.$$

We summarize the preceding results on K_{n_1,n_2}.

$$N(K_{n_1,n_2}) = \begin{cases} n_1 + n_2 - 1 & \text{for } 1 = n_1 \le n_2 \\ 2 & \text{for } n_1 = n_2 = 2 \\ n_1 + n_2 - 2 & \text{for } n_1 \equiv n_2 \equiv 2 \,(\text{mod } 3). \end{cases}$$

In general,

$$n_1 + n_2 - 2 \le N(K_{n_1,n_2}) \le n_1 + n_2 - 1.$$

6. <u>Concluding remarks</u>. Many of the results of the preceding section can also be derived from the recent interesting work of Brandenburg, Gopinath and Kurshan [1]. In particular, they establish the following theorem.

THEOREM. A graph G with n vertices can be addressed with elements of S^N if and only if there exist binary-valued n × N matrices P and Q such that $D(G) = PQ^t + QP^t$.

As previously mentioned, no counterexamples are known to the inequality $N(G) \leq n - 1$. However, this inequality has not even been established for the class of graphs G satisfying $d_G(v_i, v_j) \leq 2$ for all v_i, v_j in G. The example of $K_{2,3}$ in the preceding section shows that the stronger assertion $N(G) = \max\{n_+, n_-\}$ does not always hold.

The fact that $\det D(T_n)$ is independent of the structure of the tree T_n (cf. Eq. (8)) was initially unexpected. It is true, though, that the eigenvalues of $D(T_n)$ do depend on the structure of T_n. It seems likely that as the number of edges in a connected graph G increases the possible range of $\det D(G)$ increases, at least for a while. It would be of interest to know what this range is as a function of the number of edges of G. Perhaps $\det D(G)$ has a simple enumerative interpretation which would make these relationships clear.

References

1. L. H. Brandenburg, B. Gopinath and R. P. Kurshan, On the addressing problem of loop switching, Bell System Tech. Jour., to appear.

2. R. L. Graham and H. O. Pollak, On the addressing problem for loop switching, Bell System Tech. Jour. 50 (1971), 2495-2519.

3. B. W. Jones, The arithmetic theory of quadratic forms, CARUS Mathematical Monograph No. 10, Math. Assoc. of Amer.,1960, Providence.

4. J. R. Pierce, Network for block switching of data, Bell System Tech. Jour., to appear.

5. W. T. Trotter (personal communication).

CROSSING NUMBERS OF GRAPHS

Richard K. Guy
University of Calgary
Calgary, Alberta, Canada

A graph, $G(V,E)$, is a set V of vertices and a subset E of the
unordered pairs of (distinct) vertices, called edges. A drawing is a
mapping of a graph into a surface. The vertices go into distinct
points, called nodes. An edge and its incident vertices map into a
homeomorphic image of the closed interval $[0,1]$ with the relevant
nodes as endpoints and the interior, an arc, containing no node. A
good drawing is one in which no two arcs incident with a common node
have a common point, and no two arcs have more than one point in com-
mon. A common point of two arcs is a crossing. An optimal drawing in
a given surface is one which exhibits the least possible number of
crossings: optimal drawings are good. This least number is the
crossing number of the graph for the surface. Subdrawings of good
drawings are good, but we shall see that subdrawings of optimal
drawings are not necessarily optimal. We denote the crossing number
of G for the plane (or sphere) by $\nu(G)$. Almost all questions that
one can ask about crossing numbers remain unanswered.

For the complete graph K_n, it has been conjectured [8] that

$$(?) \qquad \nu(K_n) = [n/2]\,[(n-1)/2]\,[(n-2)/2]\,[(n-3)/2]/4, \qquad (1)$$

where brackets denote greatest integer not greater than. We establish
(1) for $n \le 10$, and indicate how it may be possible to extend this
to $n \le 12$.

To see that $\nu(K_5) = 1$, we note that Figure 1 shows it to be at
most that. We give three proofs that $\nu(K_5) > 0$. The first is by an
appeal to Kuratowski's theorem [22], which states that a graph is
planar (i.e. $\nu(G) = 0$) only if it contains no subgraph homeomorphic to
K_5 or to $K_{3,3}$ (the Thomsen, or 'gas, water, electricity' graph).
The second uses a theorem of Eggleton and Guy [3] that the parity of
the number of crossings in all good drawings of K_n is the same,
provided n is odd (i.e. if the valence is even). The third uses
Euler's formula. Suppose we can embed K_5 in the plane or sphere,
i.e., draw it without crossings. Such a drawing induces a map (in the

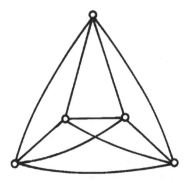

Figure 1.

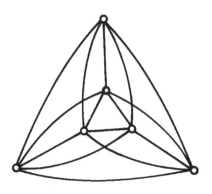

Figure 2.

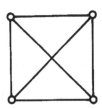

Figure 3.

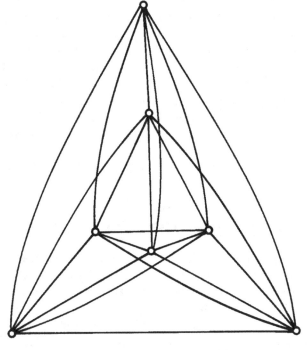

Figure 4.

cartographic sense) with 5 vertices, 10 borders and hence $10 + 2 - 5 = 7$ regions. Since each region has at least 3 borders we have $3 \times 7 \leq 2 \times 10$, a contradiction.

We will call two drawings isomorphic when there is a one-to-one correspondence between the nodes so that if any pair of arcs in one drawing crosses, the corresponding pair in the second drawing also crosses. Any optimal drawing of K_5 is isomorphic to Figure 1.

We give two proofs that $\nu(K_6) = 3$. The first uses Euler's theorem again. A drawing of K_6 with c crossings induces a map with $6 + c$ vertices, $15 + 2c$ borders and hence $11 + c$ regions. So $3(11+c) \leq 2(15+2c)$, and $c \geq 3$. That 3 crossings suffice is seen from Figure 2. For the second proof we use a special case of a general counting argument. K_m contains $\binom{m}{n}$ subgraphs K_n. Each crossing is the responsibility of just 4 nodes and so occurs in exactly $\binom{m-4}{n-4}$ such subgraphs. Thus

$$\nu(K_m) \geq \binom{m}{n} \nu(K_n) \binom{m-4}{n-4}. \tag{2}$$

In particular, if we put $m = n + 1$ we see that if (1) holds for n odd, then it also holds for $n + 1$. Since (1) is true for $n = 5$, it holds also for $n = 6$. Any optimal drawing of K_6 is isomorphic to Figure 2, since the map it induces has 14 regions, all triangles. (No triangle can have more than one crossing for a vertex, so the map consists of 3 copies of Figure 3 and two triangles. Since the graph is 5-valent, there are just two copies of Figure 3 and one triangle at each vertex, and Figure 2 is unique up to isomorphism.)

Figure 4 shows that $\nu(K_7) \leq 9$ and we know that $\nu(K_7)$ is odd. Define the responsibility of a node as the total number of crossings on all arcs incident with that node. Since each crossing is in the responsibility of 4 nodes, the total responsibility of all nodes is $4c$, where c is the number of crossings. So a drawing of K_7 contains a node with responsibility at least $\{4c/7\}$, where braces denote least integer not less than. Removal of this node and its incident edges leaves a drawing of K_6 with at most $c - \{4c/7\} = [3c/7]$ crossings. So all optimal drawings of K_7 can be obtained by introducing a new node into Figure 2. Figure 2 contains only 3 essentially different regions, and it can be verified that there are just 5 non-isomorphic optimal drawings of K_7 (Figures 4 to 8) and that $\nu(K_7) = 9$.

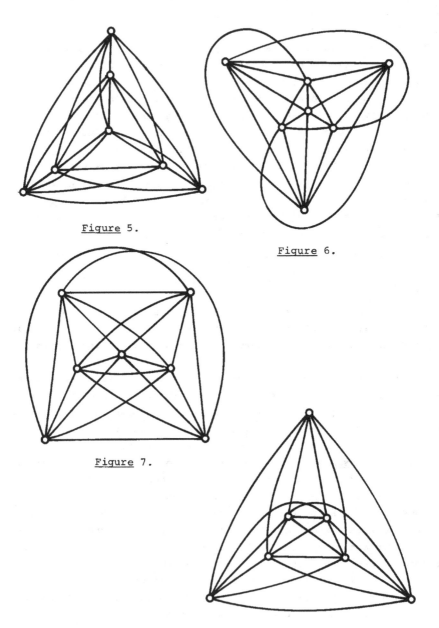

Figure 5.

Figure 6.

Figure 7.

Figure 8.

A drawing of K_8 with c crossings contains a node of responsibility $\{4c/8\}$ whose removal leaves a drawing of K_7 with $[c/2] \geq 9$, so $c \geq 18$. All drawings of K_8 with 18 crossings can be obtained by introducing a node into a region in one of Figures 4 to 8. There are just 3 non-isomorphic ways of doing this (Figures 9 - 11) and $\nu(K_8) = 18$.

The last result also follows immediately from (2). This, or the responsibility argument, and the fact that $\nu(K_9)$ is even, shows that $\nu(K_9) \geq 34$. If there is a drawing of K_9 with c crossings, then we have a subdrawing of K_8 with $[5c/9]$ crossings, so any drawing with $c = 34$ contains an optimal subdrawing of K_8. It can be verified that it is impossible to introduce a new node into any of Figures 9 to 11 and produce a drawing with 34 crossings. Figure 12 shows a drawing of K_9 with 36 crossings, so $\nu(K_9) = 36$ and (2) gives $\nu(K_{10}) = 60$. The number of non-isomorphic optimal drawings of K_9 is about 200. R. B. Eggleton hopes shortly to complete a census of these. The responsibility argument shows that the removal of any vertex from an optimal drawing of K_{10} leaves an optimal drawing of K_9, so there is a possibility that non-isomorphic optimal drawings of K_{10} are not too numerous to display.

It is at this point that the inadequacy of our methods becomes patent. Let us revert to an early tool and use Euler's formula. A drawing of K_n with c crossings induces a map with $n + c$ vertices, $\binom{n}{2} + 2c$ borders, and hence $(n^2-3n+4)/2 + c$ regions. The regions have at least 3 borders, so $3((n^2-3n+4)/2 + c \leq 2\binom{n}{2} + 4c$ and $c \geq (n-3)(n-4)/2$. We have equality for $3 \leq n \leq 6$ and not too bad a value for a few more values of n, but in general we are hopelessly far from the truth. Formula (2) shows that $\nu(K_n)/\binom{n}{4}$ is an increasing function of n. The constructions of Blažek and Koman [1] and others [e.g., 8,14] show that formula (1) provides an upper bound, so that $\nu(K_n)/\binom{n}{4}$ tends to a limit, a fact known to Erdős many years ago (written communication). This limit lies between 1/80 and 1/64. To obtain the lower bound we use a result of Kleitman [20, and see below] that $\nu(K_{5,n-5}) = 4[(n-5)/2][(n-6)/2]$, so that

$$\nu(K_n) \geq 4[(n-5)/2][(n-6)/2]\binom{n}{5}/4\binom{n-4}{3} ,$$

or

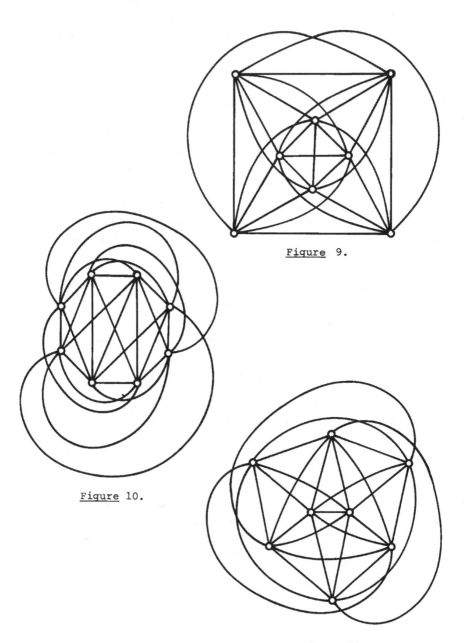

Figure 9.

Figure 10.

Figure 11.

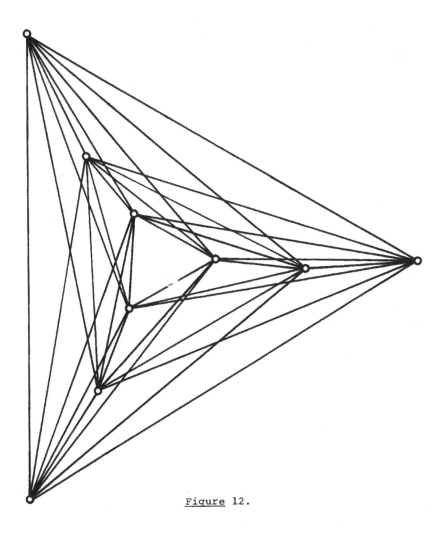

Figure 12.

(3) $\qquad \nu(K_n) \geq \left(\frac{1}{80} - \varepsilon\right) n(n-1)(n-3),$ for $n > n_0 = n_0(\varepsilon).$

Two paragraphs ago we saw that if a drawing of K_9 contains 34 crossings then it contains an optimal subdrawing of K_8. This vacuous theorem has led to a plausible, but false, conjecture: that an optimal drawing of K_n always contains an optimal subdrawing of K_{n-1}. If (1) is true, then one can demonstrate the truth of the conjecture if n is even, as we did in the same paragraph for n = 10. Regardless of the truth of (1), the conjecture is false for n odd, certainly for n = 9 and probably for all larger odd n. It is an ill wind that blows nobody any good and we can turn this perverse fact to good account by proving a conjecture of Harary and Hill [15].

If the arcs in the drawing are restricted to be straight line segments, we have the concept of the rectilinear crossing number, $\overline{\nu}(G)$, of a graph G. Rectilinear drawings are good and it is clear that $\overline{\nu}(G) \geq \nu(G)$. A theorem of Fáry [7,]7] may be stated in the form: if a graph can be embedded in the plane, then it can be so drawn using straight line segments. Hence $\nu(G) = 0$ implies that $\overline{\nu}(G) = 0$. Figures 1, 2, 4 (or 5), and 12 can each be realized with straight line segments, showing that $\nu(K_n) = \overline{\nu}(K_n)$ for $n \leq 7$ and n = 9. However, we can confirm the conjecture of Harary and Hill that $\overline{\nu}(K_8) = 19$, in contrast to $\nu(K_8) = 18$. We first note that the boundary of a drawing of K_n which is both rectilinear and optimal is a convex, crossing-free polygon. Moreover the polygon is a triangle, since if it had more sides, then its diagonals, in addition to producing one or more crossings themselves, will intersect arcs from each node interior to the polygon, contradicting optimality, since the diagonals could have been rerouted outside the polygon, with strictly less crossings. Since none of Figures 9, 10, 11 contains a crossing-free triangle, we have $\overline{\nu}(K_8) > 18$. However, we can delete a vertex of responsibility 17 (any of the three in the upper left of Figure 12) from an optimal rectilinear drawing of K_9 to leave a drawing which confirms that $\overline{\nu}(K_8) = 19$.

When the census of optimal drawings of K_9 is complete, the conjecture $\overline{\nu}(K_{10}) = 63$ can be confirmed. We add a node to each rectilinear optimal drawing of K_9. To see that this suffices, suppose that a good drawing of K_{10} contains 60 + x crossings, where $0 \leq x \leq 3$. The total responsibility is 240 + 4x, so some vertex has responsibility at least $24 + \lfloor 2x/5 \rfloor$. Since the number of crossings in the good drawing of K_9 is even, the responsibility of every vertex

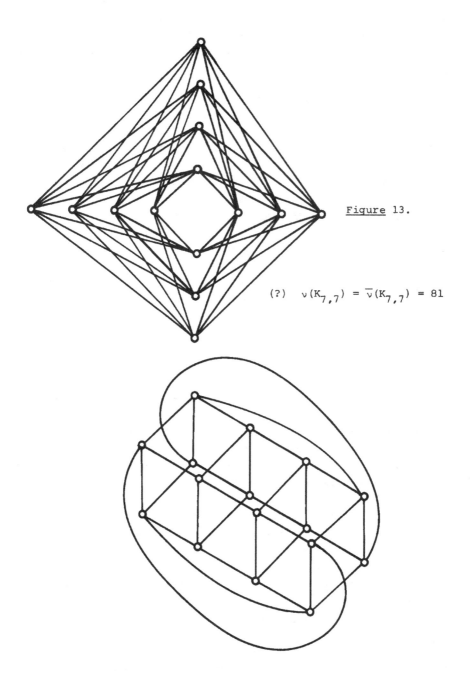

Figure 13.

(?) $\nu(K_{7,7}) = \overline{\nu}(K_{7,7}) = 81$

Figure 14.

in the good drawing of K_{10} has the same parity as x, so there is a vertex of responsibility $24 + x$, which, when deleted, leaves an optimal drawing of K_9. If the drawing of K_{10} is rectilinear, so is the subdrawing of K_9. We know that $\bar{v}(K_{10}) \leq 63$, since Jensen [16], and independently Eggleton [3], have shown that

(4) $\bar{v}(K_n) \leq [(7n^4 - 56n^3 + 128n^2 + 48n[(n-7)/3] + 108)/432]$.

In particular, add a tenth node near the bottom of the inner triangle to (a slight deformation of) Figure 12. The method will extend to show either equality in (4) for $n = 11$ and 12, i.e., that $\bar{v}(K_{11}) = 102$ and $\bar{v}(K_{12}) = 156$, or to find the correct values. It is conjectured that equality always holds in (4).

The <u>geodesic crossing number</u> of K_n on the sphere, in which minor arcs of great circles are used, is not equal to the rectilinear crossing number, since the constructions used to show that (1) is an upper bound can be made in this way on the sphere. Alternatively, Moon [23] has shown that if the nodes are taken at random on the sphere and joined by great circle arcs, then the expected number of crossings is

$$(n/2)((n-1)/2)((n-2)/2)((n-3)/2)/4,$$

so the minimum is at most this. As he points out, if airlines fly economically by flying minimum distances, then they also minimize the risk of collision!

The fact that $\bar{v}(K_5) = 1$ gives an immediate proof of Esther Klein's result [6] that five points in the plane always include a convex quadrilateral. In fact there is an exact correspondence between rectilinear crossings and convex quadrilaterals, so the problem of determining the rectilinear crossing number for the complete graph can be restated in the form: what is the least number of convex quadrilaterals determined by n points in the plane? More generally one can ask for the least number of convex k-gons determined by n points in the plane, for $k \geq 4$. Erdös observes that, as before, the ratio of this number to $\binom{n}{k}$ tends to a positive limit as n tends to infinity, for each value of k.

The crossing number problem for the complete bipartite graph $K_{m,n}$ first appeared as Turán's brick-factory problem. For some years it was thought that Zarankiewicz [30] and Urbanik [29] had solved this, but a hiatus in the proof was found independently by Ringel and Kainen [see 9], and the formula

$$(?) \qquad \nu(K_{m,n}) = [m/2]\,[(m-1)/2]\,[n/2]\,[(n-1)/2] \tag{?}$$

is still conjectural. Equality was established by Zarankiewicz for $\min(m,n) \le 3$. As before, a counting argument will give (5) for an even value of m (or n) if (5) is known to hold for the preceding (odd) number:

$$\nu(K_{m,n}) \ge \binom{m}{p}\binom{n}{q}\nu(K_{p,q})\binom{m-2}{p-2}\binom{n-2}{q-2}. \tag{6}$$

Kleitman [20] has proved equality in (5) for $\min(m,n) \le 5$ and hence for $\min(m,n) \le 6$. His result has enabled us [10] to deduce that

$$\nu(K_{m,n}) \ge m(m-1)n(n-2)/20, \tag{7}$$

provided $n \ge m \ge 5$. The smallest complete bipartite graph whose crossing number is unknown is $K_{7,7}$. The counting argument (6) and the parity argument used by Kleitman in [20] show that $\nu(K_{7,7}) = 77$, 79 or 81.

In contrast to the situation for the complete graph, it may be that the rectilinear crossing number for $K_{m,n}$ is no larger than $\nu(K_{m,n})$; it is conjectured that $\bar{\nu}(K_{m,n}) = \nu(K_{m,n})$, and is given by (5). The construction used by Zarankiewicz [30] to show that (5) is an upper bound uses only straight lines (see Figure 13).

For the 1-skelton of the n-cube, Q_n, whose vertices, the 2^n binary n-tuples, are joined by an edge just if their vertices differ in exactly one coordinate, Eggleton and Guy [4] announced that

$$(?) \qquad \nu(Q_n) \le \frac{5}{32}4^n - \left\lceil\frac{n^2+1}{2}\right\rceil 2^{n-2}, \tag{8}$$

but a gap has since been found in the description of the construction, so this must also remain a conjecture. Eggleton [3] has proved equality in (8) for $n = 4$ and has obtained all non-isomorphic optimal drawings of Q_4. We conjecture equality for all values of n.

For the complement of a circuit, $\bar{C}_n$, Guy and Hill [12] have shown that

$$\nu(\bar{C}_n) \le \frac{1}{64}(n-3)^2(n-5)^2 \qquad (n \text{ odd}), \tag{9}$$

$$\nu(\overline{C}_n) \le \frac{1}{64} n(n-4)(n-6)^2 \qquad (n \text{ even}),$$

again equality is conjectured. As in the corresponding problems for the complete graph and the complete bipartite graph, if equality holds in (9), the responsibility argument shows that (10) holds with $n + 1$ in place of n.

For more general graphs, let $G(n,k)$ be a graph with n vertices and k edges. Denote by $g(n,k)$ the minimum of $\nu(G)$ taken over all graphs $G(n,k)$. Then Erdös [5] has conjectured that

(11)   (?) $c_1 k^3/n^2 < g(n,k) < c_2 k^3/n^2$,

where c_1 and c_2 are constants, and in fact, that if $k/n \to \infty$, then $\lim(g(n,k)/(k^3/n^2))$ exists. From Euler's theorem, $g(n,3n-6) = 0$, $g(n,3n-5) = 1$. The upper bound in (11) is trivial (with $c_2 = 1/8$), for let ℓ be the least integer greater than $2k/n$ and consider n/ℓ copies of K_ℓ. The lower bound would follow if it could be proved that every drawing of a graph $G(n,k)$ contained an arc with at least $c_3 k^2/n^2$ crossings, where c_3 is a constant. Erdös also asks the following question: what is the least integer $f(r)$ so that every drawing of a graph $G(n,f(r))$ contains an arc with at least r crossings. Euler's theorem implies that $f(1) = 3n - 5$ and Eggleton and Guy [3] have shown that $f(2) = 4n - 8$ for $n = 6, 7$ or 9 and $4n - 7$ for $n = 8$ and $n \ge 10$. This implies that $g(n,k) = k - 3n + 6$ for $3n - 6 \le k \le \min(4n-8), \binom{n}{2})$, except that $g(7,20) = 6$ and $g(9,28) = 8$. The determination of $f(r)$ for $r \ge 3$ is an open problem.

Another related question is: which graphs $G(n,k)$ have maximal $\nu(G)$ and what is this maximum? Erdös conjectures that the following graph has maximal $\nu(G)$: take ℓ so that $\binom{\ell}{2} \le k < \binom{\ell+1}{2}$ and the graph consists of K_ℓ with a vertex joined to $k - \binom{\ell}{2}$ of its vertices, and $n - \ell - 1$ isolated points.

These more general problems can also be posed in the rectilinear case. We can also ask analogous questions for surfaces of higher genus. Some results have been obtained for the torus [13,14] and for projective plane and Klein bottle [21].

I am indebted to R. B. Eggleton for many helpful discussions and suggestions and for permission to reproduce his results.

References

1. J. Blažek and M. Koman, A minimal problem concerning complete plane graphs, (M. Fiedler, editor), Theory of Graphs and its Applications, Proc. Symp. Smolenice, 1963, Prague 1964, 113-117; MR 30 #4249.

2. _____, On an extremal problem concerning graphs, Comm. Math. Univ. Carolinae 8 (1967), 49-52; MR 35 #1506.

3. R. B. Eggleton, Ph.D. thesis, Univ. of Calgary, 1973.

4. R. B. Eggleton and R. K. Guy, The crossing number of the n-cube, Amer. Math. Soc. Notices 17 (1970), 757.

5. P. Erdös and R. K. Guy, Crossing number problems Amer. Math. Monthly, to appear in 1972.

6. P. Erdös and G. Szerkeres, A combinatorial problem in geometry, Compositio Math. 2 (1935), 463-470.

7. I. Fáry, On straight line representation of planar graphs, Acta Sci. Math. (Szeged) 11 (1948), 229-233; MR 10 #136.

8. R. K. Guy, A combinatorial problem, Nabla (Bull. Malayan Math. Soc.) 7 (1960), 68-72.

9. _____, The decline and fall of Zarankiewicz's theorem (F. Harary, editor), Proof Techniques in Graph Theory, Academic Press, N.Y., 1969, 63-69.

10. _____, Sequences associated with a problem of Turán and other problems, Proc. Balatonfüred Combinatorics Conference, 1969, Bolyai János Matematikai Tarsultat, Budapest, 1970, 553-569.

11. _____, Latest results on crossing numbers, Recent Trends in Graph Theory, Springer, N.Y., 1971, 143-156.

12. R. K. Guy, and A. Hill, The crossing number of the complement of a circuit, (submitted to) Discrete Math.

13. R. K. Guy and T. A. Jenkyns, The toroidal crossing number of $K_{m,n}$, J. Combinatorial Theory 6 (1969), 235-250; MR 38 #5660.

14. R. K. Guy, T. A. Jenkyns, and J. Schaer, The toroidal crossing number of the complete graph, J. Combinatorial Theory 4 (1968), 376-390; MR 36 #3682.

15. F. Harary and A. Hill, On the number of crossings in a complete graph, Proc. Edinburgh Math. Soc. (2) 13 (1962-63), 333-338; MR 29 #602.

16. H. F. Jensen, An upper bound for the rectilinear crossing number of the complete graph, J. Combinatorial Theory, Ser. B, 10 (1971), 212-216.

17. P. C. Kainen, On a problem of Erdös, J. Combinatorial Theory 5 (1968), 374-377; MR 38 #72.

18. P. C. Kainen, A lower bound for crossing numbers of graphs with applications to K_n, $K_{p,q}$ and $Q(d)$, J. Combinatorial Theory, to appear.

19. _____, On the stable crossing number of cubes, Proc. Amer. Math. Soc., to appear

20. D. J. Kleitman, The crossing number of $K_{5,n}$, J. Combinatorial Theory 9 (1970), 315-323.

21. M. Koman, On the crossing numbers of graphs, Acta Univ. Carolinae, Math. Phys. 10 (1969), 9-46.

22. K. Kuratowski, Sur le problème des courbes gauches en topologie, Fund. Math. 15 (1930), 271-283.

23. J. W. Moon, On the distribution of crossings in random complete graphs, J. Soc. Indust. App. Math. 13 (1965), 506-510; MR 31 #3357.

24. T. L. Saaty, The minimum number of intersections in complete graphs, Proc. Nat. Acad. Sci., U.S.A. 52 (1964), 688-690; MR 29 #4045.

25. _____, Two theorems on the minimum number of intersections for complete graphs, J. Combinatorial Theory 2 (1967), 571-584; MR 35 #2796.

26. _____, Symmetry and the crossing number for complete graphs, J. Res. Nat. Bureau Standards 73B (1969), 177-186.

27. W. T. Tutte, How to draw a graph, Proc. London Math. Soc. (3), 13 (1963), 743-767; MR 28 #1610.

28. _____, Towards a theory of crossing numbers, J. Combinatorial Theory 8 (1970), 45-53.

29. K. Urbaník, Solution du problème posé par P. Turán, Colloq. Math. 3 (1955), 200-201.

30. K. Zarankiewicz, On a problem of P. Turán concerning graphs, Fund. Math. 41 (1954), 137-145; Mr 16 #156.

RECENT RESULTS ON GENERALIZED
RAMSEY THEORY FOR GRAPHS*

Frank Harary
The University of Michigan
Ann Arbor, MI 48104

1. Introduction. The subject of generalized Ramsey theory for
graphs, which extended the study of Ramsey numbers from the case of
complete graphs to the investigation of arbitrary graphs (with no
isolated points), was initiated in a series of papers by Chvátal and
Harary [6,7,8,9]. Simultaneously and independently, the special case
of this generalization to the study of cycles was investigated by
Chartrand and Schuster [4,5]. The extension to directed graphs is
the topic of a forthcoming paper by Harary and Hell [16], as is the
study of Ramsey multiplicities by Harary and Prins [17]. All of these
results will be presented as well as additional observations and conjectures concerning these numbers.

2. Generalized Ramsey numbers. In this section, we follow the
development in [6]. The Ramsey number $r(m,n)$ as traditionally
studied in graph theory [14, p. 15] may be defined as the minimum number p such that every graph with p points which does not contain
the complete graph K_m must have n independent points. Alternatively, it is the smallest p for which every 2-coloring of the lines
of K_p , with two colors green and red, contains either a green K_m
or a red K_n . Thus the diagonal Ramsey numbers $r(n,n)$ can be described in terms of 2-coloring the lines of K_p and regarding K_n as
a forbidden monochromatic subgraph without regard to color.

These classical Ramsey numbers [18] involve the occurence of
monochromatic complete subgraphs in line-colored complete graphs. By
removing the completeness requirements and admitting arbitrary forbidden subgraphs within any given graph, the situation is richly and
nontrivially generalized. This viewpoint suggests the more general

    * Research supported in part by Grant 68-1515 from the Air Force
Office of Scientific Research. The author thanks A. J. Schwenk for
his assistance in the preparation of this article.

situation in which there is a c-coloring of the lines of a given graph G and the number of monochromatic occurrences of a forbidden subgraph F is calculated.

Specifically, let F have no isolates, G be a given graph, and c a positive integer. We denote by $R(G;F,c)$ the greatest integer n with the property that, in every c-coloring of the lines of G, there are at least n monochromatic occurrences of F.

Some results on these R-numbers already exist. For example, Goodman [11] found the exact value of $R(K_n;K_3, 2)$ as noted in Section 7.

In [7], the next few results are obtained. Note that $R(G;F,1)$ is the number of occurrences of F as a subgraph of G.

THEOREM. If $R(G;F,1)c^{1-q(F)} < 1$, then $R(G;F,c) = 0$.

The cube Q_n is defined as usual as the graph with 2^n points which can be taken as all binary n-sequences, with two points adjacent whenever their sequences differ in just one place.

THEOREM. The numbers $R(Q_n; Q_m, c)$ are given by the formula

$$R(Q_n; Q_m, c) = \begin{cases} \binom{n}{m}2^{n-m} & \text{if } \min(m,c) = 1, \\ \\ 0 & \text{otherwise.} \end{cases}$$

Another result, concerning R-numbers for trees, is as follows. For any two trees T_1 and T_2 such that T_1 is not a star,

$$R(T_2;T_1, c) = 0 \quad \text{whenever} \quad c \geq 2.$$

On the other hand, if $T_1 = K_{1,m}$ is a star and $d_1, d_2, \ldots, d_n$ is the degree sequence of T_2, then $R(T_2;T_1, c)$ depends only on the degree sequence of T_2, and is easily computed.

Let $r(F,c)$ be the smallest p such that every c-coloring of K_p contains a monochromatic F. Let $s = s(F)$ be the order of the automorphism group of graph F; s stands for the number of symmetries. Then it is also shown in [7] that

$$r(F,c) > (sc^{q-1})^{1/p}.$$

As immediate corollaries of this inequality, we find by substituting the exact value of s for complete bigraphs of the form $K_{n,n}$ and for cubes Q_n that

$$r(K_{n,n}, c) > c^{n/2} \left(\frac{2n!^2}{c} \right)^{1/2n},$$

$$r(Q_n, c) > c^{n/2} \left(\frac{n! 2^n}{c} \right)^{1/2n}.$$

By definition, the <u>Ramsey</u> <u>number</u>, $r(F_1, F_2)$, is the minimum p such that every 2-coloring of the lines of K_p contains a green F_1 or a red F_2 . For brevity we write $r(F) = r(F,F)$. The following conjecture, proposed in [6], has just been disproved by F. Galvin.

CONJECTURE. <u>For</u> <u>any</u> <u>two</u> <u>graphs</u> F_1 <u>and</u> F_2 <u>with</u> <u>no</u> <u>isolates</u>,

$$r(F_1, F_2) \geq \min\{r(F_1), r(F_2)\}.$$

<u>GALVIN'S</u> <u>COUNTEREXAMPLE</u>. <u>The</u> <u>graphs</u> P_5 <u>and</u> $K_{1,3}$ <u>provide</u> <u>a</u> <u>counterexample</u> <u>to</u> <u>the</u> <u>above</u> <u>conjecture</u>, <u>since</u>

$$r(P_5, K_{1,3}) = 5 < 6 = r(P_5) = r(K_{1,3}).$$

All of the numbers $r(F_1, F_2)$ have now been determined for the graphs F_i with at most four points having no isolates. Very few Ramsey numbers have been calculated for graphs with more than five points except for three simple families: The stars $K_{1,n}$ are essentially trivial since they merely involve the degrees of the points, the paths P_n and more generally the Ramsey numbers $r(P_m, P_n)$ (a formula for these numbers has been found for all m and n), and finally and least tractable the cycles C_n for which only partial results have been obtained to date. The Ramsey numbers for stars, paths, and cycles are studied in the next section and those for small graphs in Section 4.

3. <u>Complete</u> <u>graphs</u>, <u>stars</u>, <u>paths</u>, <u>cycles</u>. We first review the status of the classical Ramsey numbers for pairs of complete graphs. We then turn to those classes of graphs for which the generalized Ramsey numbers have been more or less obtained. These include stars, paths, and cycles.

The determination of Ramsey numbers is usually exceedingly difficult. The first problem suggested by Ramsey's Theorem is to determine $r(K_m, K_n)$. In 40 years, only six of these have been found with $m, n \geq 3$; they are listed in [14, p. 17]:

$$r(K_3, K_3) = 6, \quad r(K_3, K_4) = 9, \quad r(K_3, K_5) = 14,$$

$$r(K_3, K_6) = 18, \quad r(K_3, K_7) = 23, \quad r(K_4, K_4) = 18.$$

Several lower and upper bounds have been found, but no other exact values are known.

The generalization to graphs other than complete graphs provides some hope for progress and new insight. The stars are the easiest family of graphs to handle; see [3].

LEMMA. <u>The Ramsey numbers for stars are given by the formula</u>:

$$r(K_{1,m}, K_{1,n}) = \begin{cases} m + n & \text{if } m \text{ or } n \text{ is odd}, \\ m + n - 1 & \text{if } m \text{ and } n \text{ are both even}. \end{cases}$$

Proof. Consider coloring the lines of K_{m+n} red and green. At each point, there are $m + n - 1$ incident lines to be colored, so, if less than m are red, then at least n are green. Thus $r(K_{1,m}, K_{1,n}) \leq m + n$.

Case 1. If m or n is odd, say m is odd without loss of generality, then $m - 1$ is even so there must exist a regular graph of degree $m - 1$ on $m + n - 1$ points. Moreover, the complementary graph is regular of degree $n - 1$, so we see we can 2-color K_{m+n-1} with no red $K_{1,m}$ and no green $K_{1,n}$. Thus $r(K_{1,m}, K_{1,n}) = m + n$ in this case.

Case 2. If m and n are both even, we observe that there is no regular graph of degree $m - 1$ on $m + n - 1$ points, because that would comprise an odd number of points of odd degree. Thus $r(K_{1,m}, K_{1,n}) \leq m + n - 1$. We can easily find a regular graph of degree $m - 1$ on $m + n - 2$ points, so we have $r(K_{1,m}, K_{1,n}) = m + n - 1$ which completes the proof.

In particular, $r(K_{1,n}) = 2n$ if n is odd; $2n - 1$ if n is even.

Burr and Roberts [2] calculated the Ramsey numbers for paths.

LEMMA. The Ramsey numbers for paths are given by the formula:

$$r(P_m, P_n) = n - 1 + \left\lfloor \frac{m}{2} \right\rfloor \quad \text{for} \quad 2 \leq m \leq n.$$

Chartrand and Schuster have studied the Ramsey numbers of cycles. Between them [4,5] and others [1], some results have been established and others conjectured.

The cyclic Ramsey numbers $c(m,n) = r(C_m, C_n)$ are defined for $m \leq n$.

THEOREM. 1). $c(3,3) = c(4,4) = 6$ and $c(4,5) = 7$.

2). If m is odd, $(m,n) \neq (3,3)$, then $c(m,n) = 2n - 1$.

CONJECTURE. 3). If m is even, n is even, $(m,n) \neq (4,4)$, then $c(m,n) = n - 1 + \frac{m}{2}$.

4). If m is even, n is odd, $(m,n) \neq (4,5)$, then $c(m,n) = \max\left\{ 2m - 1, \ n - 1 + \frac{m}{2} \right\}$.

The present status of the proof of statement 2) is as follows. Schuster [19] established the lower bound which is of course easier: Take K_{2n-2} and color green the lines of $K_{n-1,n-1}$. The upper bound was reported to have been established by one Mrs. Rosta in a letter from the omnipresent P. Erdös to S. Schuster received at the Kalamazoo meeting. (In particular, Bondy and Erdös [1] proved that $c(n,n) = 2n - 1$ if n is odd.)

In Conjectures 3) and 4), the lower bound is similarly easy to establish (using $K_{n-1,m/2-1}$ and $K_{m-1,m-1}$); the upper bound is the rub. Nevertheless it appears that these conjectures are true and will soon be settled. For further information on cyclic Ramsey numbers, see Schuster [19].

The generalized Ramsey numbers (F_1, F_2) obtained above for stars, paths, and cycles all involve the confrontation of two graphs F_1 and F_2 from the same family. However, the particular problem of comparing a star with a complete graph is relatively simple to solve and this has been done by A. J. Schwenk.

THEOREM. <u>The Ramsey number</u> $r(K_m, K_{1,n})$ <u>is given by the</u>
<u>formula</u>:

$$r(K_m, K_{1,n}) = (m - 1)n + 1.$$

Proof. To get the lower bound, we color red the lines of the
complete $(m-1)$-partite graph $K_{n,n,\ldots,n}$ (see [14, p. 23]). Clearly
there is no red K_m, and the complement is $(m-1)K_n$ which is regular of degree $n - 1$, and so has no green $K_{1,n}$.

The upper bound is obtained by induction on m. When $m = 2$,
the bound is trivially $n + 1$. Suppose $K_{(m-1)n+1} = G \cup \overline{G}$ represents a 2-coloring of $K_{(m-1)n+1}$. Now if $\overline{G}$ has no $K_{1,n}$, we conclude G has minimum degree at least $(m - 2)n + 1$. Select a point
v, and observe that its neighborhood has size at least $r(K_{m-1}, K_{1,n})$.
Thus, since $\overline{G}$ does not contain $K_{1,n}$, the neighborhood of v must
contain K_{m-1}. Together with v this provides a monochromatic K_m,
completing the proof.

The upper bound half of the proof can also be obtained as a
corollary to Turán's theorem [14, p. 18]. Namely, in a graph G with
$p = (m - 1)n + 1$ and minimum degree $(m - 2)n + 1$ (so that its
complement has maximum degree $n - 1$), the number q of lines is
bounded by

$$q \geq \frac{(mn - n+1)(mn - 2n+1)}{2} > \frac{n(m-2)(mn - n+2)}{2}.$$

Then, according to Turán's theorem, G must contain K_m.

4. <u>Ramsey numbers for small graphs</u>. There are exactly ten
graphs F (Figure 1) with at most 4 points, having no isolates.
For convenience in identifying them, we use the operations on graphs
from [14, p. 21], to write a symbolic name for each.

Table 1 summarizes the results we have obtained for the 10
small graphs shown in Figure 1.

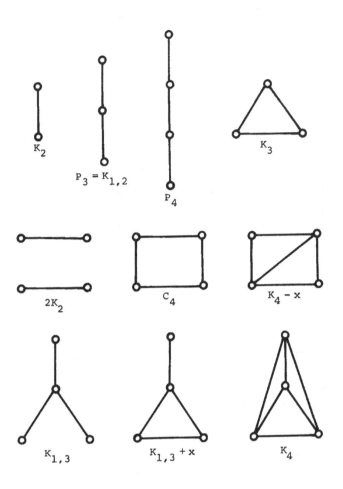

Figure 1.

    The proofs of the diagonal entries in Table 1 are given in [8] and of the off-diagonal entries in [9]. The general method is illustrated by demonstrating the one entry $r(K_{1,3} + x) = 7$. There are as usual two steps in proving that $a = b$, namely verifying both $a \geq b$ and $a \leq b$. The fact that $r(K_{1,3} + x) \geq 7$ follows from a 2-coloring of K_6 in which $K_{3,3}$ (containing no triangles) is colored green and its complement $2K_3$ (containing no point of degree 3) is colored red.

	K_2	P_3	$2K_2$	K_3	P_4	$K_{1,3}$	C_4	$K_{1,3}+x$	K_4-x	K_4
K_2	2	3	4	3	4	4	4	4	4	4
P_3		3	4	5	4	5	4	5	5	7
$2K_2$			5	5	5	5	5	5	5	6
K_3				6	7	7	7	7	7	9
P_4					5	5	5	7	7	10
$K_{1,3}$						6	6	7	7	10
C_4							6	7	7	10
$K_{1,3}+x$								7	7	10
K_4-x									10	11
K_4										18

Table 1. Small generalized Ramsey numbers $r(F_1, F_2)$

To show that $r(K_{1,3} + x) \leq 7$, let us consider any 2-coloring of K_7. Since $r(K_3) = 6$, we know it must contain a monochromatic triangle, which we take to be red. Then, if we are to avoid forming a red $K_{1,3} + x$, we must color green all lines joining the 3 points of this red triangle to the remaining 4 points, thus forming a green $K_{3,4}$. Now if any line joining two points from the set of four independent points in $K_{3,4}$ is colored green, we find that a green $K_{1,3} + x$ has been created. But, on the other hand, if all such lines are colored red, we form a red K_4 which contains $K_{1,3} + x$. In other words, any 2-coloring of K_7 must contain a monochromatic $K_{1,3} + x$, and so $r(K_{1,3} + x) \leq 7$. This completes the proof that $r(K_{1,3} + x) = 7$.

5. Ramsey multiplicities. It was shown in Goodman [11] that every 2-coloring of K_6 contains at least two monochromatic triangles. In [15], we described precisely those 2-colorings of K_6 which accomplish this and proved that the two monochromatic triangles have exactly one common point if and only if they have different colors. This suggested to Schwenk [20] a shorter proof of Goodman's result [11] giving the minimum number of monochromatic triangles in a 2-coloring of K_p.

Later we found that every 2-coloring of K_6 has at least monochromatic quadrilaterals (C_4) and that two is the smallest sible number. These "data" suggested the following invariant of a graph F with no isolates. The <u>Ramsey multiplicity</u> of F, written R(F), is the smallest possible number of monochromatic occurrences of F in any 2-coloring of $K_{r(F)}$.

The Ramsey multiplicities of 8 of the 10 small graphs shown in Figure 1 have been determined by Harary and Prins [17]. These values together with one conjecture (for $K_4 - x$) and one question mark (for K_4) are:

F	K_2	P_3	K_3	$2K_2$	P_4	$K_{1,3}$	C_4	$K_{1,3}+x$	K_4-x	K_4
r(F)	2	3	6	5	5	6	6	7	10	18
R(F)	1	1	2	3	10	3	2	12	15?	?

<u>Table</u> 2. Ramsey multiplicities

The Ramsey multiplicities of all stars have been determined:

$$R(K_{1,n}) = \begin{cases} 1 & n \quad \text{even,} \\ n & n \quad \text{odd.} \end{cases}$$

These considerations lead to the following inference which appears thus far to be true.

CONJECTURE. <u>The</u> <u>Ramsey</u> <u>multiplicity</u> <u>of</u> <u>a</u> <u>graph</u> F <u>with</u> <u>no</u> <u>isolates</u> <u>is</u> 1 <u>if</u> <u>and</u> <u>only</u> <u>if</u> F <u>is</u> K_2 <u>or</u> F <u>is</u> <u>a</u> <u>star</u> $K_{1,n}$ <u>with</u> n <u>even</u>.

6. <u>Ramsey games</u>. The necessity of the occurrence of a monochromatic forbidden subgraph F in any 2-coloring of K_p when p = r(F,F) immediately suggests a multitude of possible two-person games. For example, if two players, Red and Green, alternate coloring the lines of K_6 with K_3 as the forbidden subgraph, we may define (at least) four possible games:

1) <u>Avoidance Game</u>: the first player to form a monochromatic triangle of his color loses.

2) __Achievement__ Game: the first player to form a monochromatic triangle wins.

3) __Minimization__ Game: the game proceeds until all lines are colored. Then the player with fewer triangles wins.

4) __Maximization__ Game: the player who forms the greater number of triangles wins.

Note that the first two variations cannot end in a draw, since a monochromatic triangle must occur when all the lines are colored. However, these two versions seem to be more easily analyzed than the other two, so Games 3) and 4) are preferred. Game 3) has an antagonistic spirit in that one tries to force his opponent into forming a triangle. Thus, Game 4) was chosen when this talk was presented to illustrate this type of game. The genial director of this conference, Prof. Y. Alavi, amply represented the novices and did not disgrace himself while playing the author to a 1 to 1 draw, which gave the smallest possible total score.

In general, these games do not seem to yield to any strategy other than exhaustive analysis. However, it appears that one can do quite well playing on K_6 with $F = K_3$ if one repeatedly adjusts one of the color classes to approximate $K_{3,3}$, that is, temporarily imagine the points being partitioned into two sets of three. Clearly the choice will depend upon which of these four games is being played.

It may well be that this is the sort of two person game which has been long sought by students of artificial intelligence because there is some promise that an appropriate computer program will enable a computer to beat his human opponent rather consistently. Actually, G.J. Simmons proved that in the Avoidance Game, Player 2 has a guaranteed win.

7. __Colorings__ __of__ __noncomplete__ __graphs__. Another variation of Ramsey numbers is obtained when we remove the restriction that we are coloring the lines of complete graphs. We define the __Ramsey__ __coloring__ __number__ __of__ G __with__ __respect__ __to__ F_1 __and__ F_2, written $R(G; F_1, F_2)$ as the minimum sum among all 2-colorings of G of the number of red F_1's plus the number of green F_2's. In terms of this notation, the Ramsey multiplicity becomes

$$R(F_1, F_2) = R(K_p; F_1, F_2) \quad \text{where} \quad p = r(F_1, F_2).$$

Goodman [11] was the first to determine Ramsey coloring numbers when he found $R(K_p; K_3, K_3)$. His result, as it appears in Schwenk [20], is

THEOREM. The Ramsey coloring number of K_p with respect to K_3 is given by the formula:

$$R(K_p; K_3, K_3) = \binom{p}{3} - \left[\frac{p}{2}\left[\left(\frac{p-1}{2}\right)^2\right]\right].$$

Inasmuch as $r(K_3, K_3) = 6$, Erdős once asked if $R(G; K_3, K_3) > 0$ implies $G \supset K_6$. Graham [12] answered this query in the negative by showing that $K_8 - C_5$ is the unique smallest graph G not containing K_6 for which $R(G; K_3, K_3) > 0$. (Curiously, this is also the unique smallest example of a maximal toroidal graph which is not a triangulation, as shown in a forthcoming joint paper by Harary, Kainen, Schwenk, and White.)

Next, Graham and Spencer [13] found a graph G with 23 points and no K_5 for which $R(G; K_3, K_3) > 0$. Finally, Folkman [10] proved that there exists a graph with no K_4 for which $R(G; K_3, K_3) > 0!$ Unfortunately, his argument only proves the existence of such a graph with some astronomical number of points. No one has yet found a graph with no K_4 on a relatively small number of points.

8. Ramsey theory for digraphs. We may extend Ramsey theory to digraphs, following Harary and Hell [16], by letting $\vec{K}_p$ denote the complete digraph with a symmetric pair of arcs joining each pair of its p points. Then $r(D_1, D_2)$ is defined as the smallest p such that any 2-coloring of (the arcs of) $\vec{K}_p$ has a red D_1 or a green D_2. Ramsey's original theorem says Ramsey numbers exist for graphs. Surprisingly, they don't always exist for digraphs!

THEOREM 1. The Ramsey number $r(D_1, D_2)$ exists if and only if at least one of D_1 or D_2 is acyclic.

To see the necessity, we 2-color $\vec{K}_p$ so that each color class is acyclic. This is accomplished by coloring red the arcs of the transitive tournament T_p. The remaining arcs, $\overline{T}_p$, also form a transitive tournament, with the order of the points reversed. Thus we see that $\vec{K}_p$ can be 2-colored so that both color classes are acyclic. Consequently, if D_1 and D_2 both have directed cycles, neither need ever occur upon 2-coloring $\vec{K}_p$ for any p.

Occasionally one is tempted to discount Ramsey's theorem as being intuitively obvious. This result for digraphs shows that such a number need not exist for all types of graphical structure.

Let GD be the underlying graph of the digraph D, with the same point set and with u,v adjacent in GD whenever u,v are joined by 1 or 2 arcs in D. We can bound Ramsey numbers for digraphs by the corresponding Ramsey numbers for graphs.

THEOREM 2. For digraphs D_1 and D_2,

$$r(D_1 , D_2) \geq r(GD_1 , GD_2).$$

Proof. A 2-coloring of K_p with $p = r(GD_1 , GD_2) - 1$ avoiding GD_1 and GD_2 gives a 2-coloring of $\vec{K}_p$ avoiding D_1 and D_2 if we color both arcs uv and vu in the digraph the same color as line uv in the graphical coloring.

Another useful bound is given in Theorem 3. Its importance is enhanced by Theorem 4, equating Ramsey numbers for transitive tournaments to the traditional Ramsey numbers for complete graphs.

THEOREM 3. If D_1 and D_2 are acyclic digraphs with m and n points respectively, then

$$r(D_1 , D_2) \leq r(T_m , T_n).$$

Proof. Every acyclic digraph can be embedded in the transitive tournament on the same number of points (see [14, p. 200]). Thus, since every 2-coloring of $\vec{K}_p$ with $p = r(T_m , T_n)$ contains a red T_m or a green T_n, a fortiori, every coloring contains a red D_1 or a green D_2.

THEOREM 4. The Ramsey numbers for transitive tournaments and complete graphs are related by the formula:

$$r(T_m , T_n) = r(K_m , K_n) = r(m,n).$$

Here the second equality merely presents two notations for the traditional Ramsey numbers. The fact that $r(T_m , T_n) \geq r(m,n)$ follows immediately from Theorem 2. The opposite inequality is proved in Harary and Hell [16].

We close with a result which has recently been proved by J. Williamson [21].

Let $\vec{P}_m$ denote the directed path with m points.

THEOREM 5. <u>The</u> <u>Ramsey</u> <u>numbers</u> <u>for</u> <u>directed</u> <u>paths</u> <u>are</u> <u>given</u> <u>by</u> <u>the</u> <u>formula</u>:

$$r(\vec{P}_2, \vec{P}_n) = n \quad \underline{for} \quad 2 \le n,$$

$$r(\vec{P}_m, \vec{P}_n) = n + m - 3 \quad \underline{for} \quad 3 \le m \le n.$$

9. Conclusion. Virtually all of the known results on generalized Ramsey theory for graphs have been reported here, and the most general method of proof was brute force. There is certainly a need for more powerful and general methods, but it is not certain that these exist. Since the study of Ramsey properties of general graphs appears less intractable than that for complete graphs, this may well suggest fruitful directions for other mathematical structures such as vector spaces. The fact that generalized Ramsey theory for graphs is in its infancy is attested by more than half of the references having the status, "to appear".

References

1. J. A. Bondy and P. Erdös, Ramsey numbers for cycles in graphs, to appear.

2. S. A. Burr and J. A. Roberts, On Ramsey numbers for paths, to appear.

3. _____, On Ramsey numbers for stars, to appear.

4. G. Chartrand and S. Schuster, On the existence of specified cycles in complementary graphs, <u>Bull</u>. <u>Amer</u>. <u>Math</u>. <u>Soc</u>. 77 (1971), 995-998.

5. _____, A note on cycle Ramsey numbers, <u>Discrete</u> <u>Math</u>., submitted for publication.

6. V. Chvátal and F. Harary, Generalized Ramsey theory for graphs, <u>Bull</u>. <u>Amer</u>. <u>Math</u>. <u>Soc</u>., to appear.

V. Chvátal and F. Harary, Generalized Ramsey theory for graphs, I, diagonal numbers, *Periodica* Math. *Hungar*., to appear.

_____, Generalized Ramsey theory for graphs, II, small diagonal numbers, *Proc*. *Amer*. *Math*. *Soc*. 32 (1972), 389-394.

_____, Generalized Ramsey theory for graphs, III, small off-diagonal numbers, *Pacific* J. *Math*., to appear.

10. J. Folkman, Graphs with nonochromatic complete subgraphs in every edge coloring, *Studies* in *Combinatorics* (J. B. Kruskel and E. C. Posner, eds.), SIAM, Philadelphia, 1970.

11. A. W. Goodman, On sets of acquaintances and strangers at any party, *Amer*. *Math*. *Monthly* 66 (1959), 778-783.

12. R. L. Graham, On edgewise 2-colored graphs with monochromatic triangles and containing no complete hexagon, J. *Combinatorial Theory* 4 (1968), 300.

13. R. L. Graham and J. Spencer, On small graphs with forced monochromatic triangles, *Recent* *Trends* in *Graph* *Theory*, (M. Capobianco, et al., eds.), Springer, New York, 1971, 137-141.

14. F. Harary, *Graph* *Theory*, Addison-Wesley, Reading, Mass., 1969.

15. F. Harary, The two triangle case of the acquaintance graph, *Math*. *Mag*. 45 (1972), 130-135.

16. F. Harary and P. Hell, Generalized Ramsey theory for graphs, IV, Ramsey numbers for digraphs, to appear.

17. F. Harary and G. Prins, Generalized Ramsey theory for graphs, V, Ramsey multiplicities, to appear.

18. F. P. Ramsey, On a problem of formal logic, *Proc*. *London* *Math*. *Soc*. 30 (1930), 264-286.

19. S. Schuster, Progress on the problem of eccentric hosts, this volume, 283-290.

20. A. Schwenk, The acquaintance graph revisited, *Amer*. *Math*. *Monthly*, to appear.

21. J. Williamson, A Ramsey-type problem for paths in digraphs, to appear.

LINE GRAPHS OF TRIANGLELESS GRAPHS
AND ITERATED CLIQUE GRAPHS*

S. T. Hedetniemi
University of Iowa
Iowa City, IA 22903
and
University of Virginia
Charlottesville, VA 22903

and

P. J. Slater
University of Iowa
Iowa City, IA 52240

Perhaps no other general class of graphs has proved to be so amenable to characterizations as the class and subclasses of line graphs. Line graphs have been characterized in at least three different ways (cf. Krausz [8], van Rooij and Wilf [10], and Beineke [1]) including a forbidden subgraph characterization. Line graphs of complete graphs have been characterized by Chang [2], Hoffman [6], and Turri [11]. Line graphs of complete bipartite and regular bipartite graphs have been characterized by Moon [9], Hoffman [7], and Turri [11]. Line graphs of all bipartite graphs have been characterized in four different ways by Hedetniemi [5], including a forbidden subgraph characterization. And finally, line graphs of trees have been characterized by Chartrand and Turri [cf. 4, 11].

So often in these characterizations, it is the form of one characterization of one subclass of graphs that suggests the form of a characterization of another subclass. The results in this paper grew directly out of results found in [5] and [3]. We extend the characterization in [5] of line graphs of bipartite graphs to line graphs of graphs having no triangles.

In addition, the form of this new characterization, which involves the use of clique graphs, strongly suggests how the development of a study of iterated graphs will go. We also present several results in this direction.

---

* This research was supported in part by Grant number N00014-68-A-0500, Office of Naval Research.

We will first need a few basic definitions; terms not defined here are defined in Harary [4]. If $G = (V,E)$ is a graph, then the <u>line graph</u> $L(G)$ <u>of</u> $\underline{G}$ is the graph $L(G) = (V',E')$, whose set of points V' corresponds $1 - 1$ with the set of lines E of G, and two points u' and v' of $L(G)$ are adjacent, i.e., $u'v' \in E'$, if and only if the two lines of E to which these two points correspond have a point of G in common.

A <u>clique</u> C of a graph $G = (V,E)$ is a maximal complete subgraph of G. The <u>clique graph</u> $K(G)$ <u>of a graph</u> $G = (V,E)$ is the graph $K(G) = (V',E')$, whose set of points V' corresponds $1 - 1$ with the set of distinct cliques of G, and two points u',v' are adjacent in $K(G)$, i.e., $u'v' \in E'$ if and only if the two cliques of G to which these two points correspond have a point in common. By $K^2(G)$ we mean $K(K(G))$, and in general $K^n(G) = K(K^{n-1}(G))$.

In Figure 1 we present an example of a graph G, its line graph $L(G)$ and its clique graph $K(G)$.

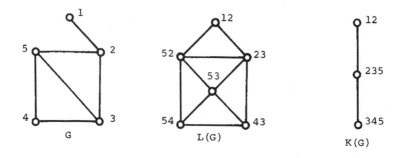

G   L(G)   K(G)

Figure 1.

We will have occasion later on to refer to two specific graphs, $K_4 - x$ and $K_{1,3}$; these two graphs are shown in Figure 2.

We are now ready for our first result.

THEOREM 1. If G <u>is the</u> <u>line graph of a graph</u> H <u>which has no triangles then no two cliques</u> C_i <u>and</u> C_j <u>of</u> G <u>have more than one point in common, i.e.,</u> $|C_i \cap C_j| \leq 1$, <u>and</u> $K(G)$ <u>has no triangles</u>.

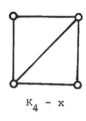

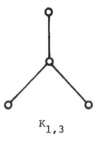

$$K_4 - x \qquad\qquad K_{1,3}$$

Figure 2.

Proof. Let H be a graph having no triangles and let G = L(H). Let $C_1, C_2, \ldots, C_n$ be the cliques of G. Suppose there exist two cliques C_i and C_j such that $|C_i \cap C_j| > 1$. Then G must contain a line uv which belongs to cliques C_i and C_j. Since C_i and C_j are distinct cliques there must exist a point $w \in C_i$ which is not in C_j and there must exist a point $x \in C_j$ which is not in C_i and furthermore w and x are not adjacent. Therefore G must contain a subgraph isomorphic to $K_4 - x$ as in Figure 3, where $u,v,w \in C_i$

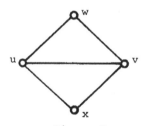

Figure 3.

and $u,v,x \in C_j$. Now G is the line graph of H; hence every point of G corresponds to a line of H. Let point u correspond to the line $u_2 u_3$ of H. Then point x corresponds to a line which is incident with either u_2 or u_3, say u_2. Let x correspond to a line $u_2 u_1$ in H.

Next, point w corresponds to a line in H which is adjacent to line $u_2 u_3$ but is not adjacent to line $u_2 u_1$. Hence w must correspond to a line, say $u_3 u_4$, $u_4 \neq u_1$. Finally, point v must correspond to a line in H which is incident with all three lines $u_2 u_3$, $u_2 u_1$ and $u_3 u_4$, i.e., the line corresponding to point v must contain u_2 or u_3, and u_2 or u_1, and u_3 or u_4. There are only two possibilities; this line is either u_1u_3 or u_2u_4. But in either case it would follow that H has a triangle; a contradiction. Hence $|C_i \cap C_j| \le 1$ for any two cliques of G.

We must now show that $K(G)$ has no triangles. Suppose that $K(G)$ has a triangle, formed by cliques C_i, C_j and C_k.

Since we have just shown that $|C_i \cap C_j| \le 1$ for any two cliques of G, we have two cases to consider as to how this triangle of cliques is formed:

Case 1. There exists in G one point u which belongs to all three cliques, i.e., $u \in C_i$, $u \in C_j$, and $u \in C_k$. Now since all three cliques are distinct, there must exist points $u_i \in C_i$, $u_j \in C_j$, and $u_k \in C_k$, each of which is not an element of the other two cliques, because we know $|C_i \cap C_j| \le 1$.

Consider then the subgraph of G induced by the four points u, u_i, u_j, u_k, where we know that uu_i, uu_j and uu_k must be lines of G. These three lines alone induce the subgraph $K_{1,3}$. But G is a line graph, and this is a forbidden subgraph of any line graph (cf. Beineke [1]). Hence at least one of the lines u_iu_j, u_ju_k or u_ku_i must exist in G. This extra line however would form a triangle with point u in G and hence G would contain yet another clique C, containing this triangle, $C \ne C_i$, $C \ne C_j$, and $C \ne C_k$. But this would imply that either $|C \cap C_i| \ge 2$, or $|C \cap C_j| \ge 2$ or $|C \cap C_k| \ge 2$, which contradicts our assumption that $|C_i \cap C_j| \le 1$ for any two cliques of G.

Case 2. The triangle of cliques is formed by $C_i \cap C_j = \{u_1\}$, $C_j \cap C_k = \{u_2\}$ and $C_k \cap C_i = \{u_3\}$, where all three points u_1, u_2, and u_3 are distinct. But now we see that since u_1 and u_2 are both in C_j, u_1u_2 is an edge of G. Similarly, u_2u_3 and u_3u_1 must be edges of G. Thus u_1, u_2 and u_3 form a triangle in G, and by definition this triangle must be contained in sone clique C of G, where $C \ne C_i$, $C \ne C_j$ and $C \ne C_k$. But this contradicts our assumption that $|C_i \cap C_j| \ge 2$. Thus Cases 1 and 2, assuming that $K(G)$ has a triangle, both lead to situations which contradict our assumption that $|C_i \cap C_j| \le 1$ for any two cliques of G. Thus $K(G)$ has no triangles.

We would now like to consider a possible converse to Theorem 1: Does it follow that if a graph G satisfies the condition, $|C_i \cap C_j| \le 1$ for any two cliques of G, and the condition, $K(G)$

has no triangles, then G is isomorphic to the line graph of a graph
H which has no triangles? Our next result shows that the converse
does indeed hold.

THEOREM 2. <u>If</u> G <u>is a graph which satisfies the condition</u>
$|C_i \cap C_j| \leq 1$ <u>for any two cliques</u> C_i, C_j <u>of</u> G <u>and satisfies the</u>
<u>condition that the clique graph</u> K(G) <u>has no triangles, then</u> G <u>is</u>
<u>isomorphic to the line graph of a graph</u> H <u>which has no triangles.</u>

<u>Proof.</u> Assume, without loss of generality, that G is connected,
K(G) is triangle-less and $|C_i \cap C_j| \leq 1$ for any two cliques of G.
Let C_1, C_2, ..., C_n be the cliques of G. Since $|C_i \cap C_j| \leq 1$ for
any two cliques of G, each line of G lies in exactly one clique.
Therefore C_1, C_2, ..., C_n determines a partition of the lines of G
into complete subgraphs (every line belongs to at least one clique).
Furthermore, no point of G lies in more than two of these complete
subgraphs, for if one point were in three of these cliques C_i, C_j
and C_k then these three cliques would form a triangle in K(G),
which would contradict our assumption that K(G) has no triangles.
Thus, by Krausz's characterization of line graphs [8], it follows that
G is the line graph of some graph H.

We must show that H has no triangles. Let H have a triangle
v_1v_2, v_2v_3, v_3v_1. Then in G = L(H) there is a triangle formed by
the three points corresponding to these three lines; see Figure 4

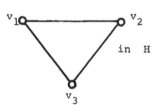
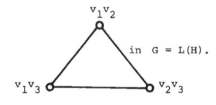

Figure 4.

Now, if H has another line, then since G is assumed to be connected, there must exist a point w in H which is adjacent to at
least one of v_1, v_2 or v_3. Assume that w is adjacent to v_1,
i.e., $wv_1 \in E(H)$. Then in G, the point corresponding to wv_1 is
adjacent to both v_1v_2 and v_1v_3, but is not adjacent to v_2v_3; see
Figure 5.

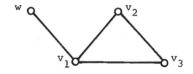 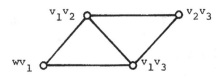

Figure 5.

Since wv_1 is not adjacent to v_2v_3, the two triangles wv_1, v_1v_2, v_1v_3 and v_1v_2, v_2v_3, v_1v_3 belong to two different cliques C_1, C_j of G. But $|C_i \cap C_j| \geq 2$ which contradicts our assumption that $|C_i \cap C_j| \leq 1$.

Therefore if H has a triangle v_1v_2, v_2v_3, v_1v_3 then H cannot have other lines. But this means that $G = L(H) = K_3$. But K_3 also happens to be the line graph of $K_{1,3}$, a graph having no triangles.

Thus, either H has no triangles, or else H has exactly one, in which case $G = K_3$ which is the line-graph of $K_{1,3}$, a graph with no triangles.

We have thus observed by Theorems 1 and 2 that <u>a graph</u> G <u>is the line graph of a graph</u> H <u>without triangles if and only if</u> $|C_i \cap C_j| \leq 1$ <u>holds for any two cliques of</u> G <u>and</u> $K(G)$ <u>has no triangles</u>.

We would now like to establish a result that states that the graph H above, having no triangles, and the graph $K(G)$, having no triangles, are nearly the same.

THEOREM 3. <u>If</u> H <u>is a connected graph containing no triangles and at least three points, then</u> $K(L(H)) = K^2(H) \cong H - \{u \mid \deg u = 1\}$.

<u>Proof</u>. Let H have points $u_1u_2 \cdots u_m$, u_{m+1}, $\ldots$, u_p, where u_{m+1}, $\ldots$, u_p are the points of H having degree one. Since H is assumed to be connected, H has no points of degree zero, unless of course $H = K_1$, a trivial case to consider.

We will now show that there is a unique clique C_i in $L(H)$ corresponding to each point u_i, $i = 1, \ldots, m$ (where $\deg u_i > 1$) and further that these are the only cliques of $L(H)$. We complete the proof by showing that $C_i \cap C_j \neq \emptyset$ (and thus $|C_i \cap C_j| = 1$) if and only if the corresponding points u_i and u_j are adjacent in H.

We note first that since H has no triangles and no points of degree zero, every clique of H is a line, i.e., a K_2, and every line of H is a clique.

Let point u_i in H be adjacent to points w_1, w_2, ..., w_n where $n \geq 2$. Then the points in $L(H)$ corresponding to the lines $u_i w_1$, $u_i w_2$, ..., $u_i w_n$ in H are all mutually adjacent and hence form a complete graph on n points in $L(H)$. We claim that this complete graph is a clique of $L(H)$, since any other point of $L(H)$ adjacent to both $u_i w_1$ and $u_i w_2$ must correspond to a line in H incident with u_i, i.e., a line of the form $u_i w_j$. Call this clique of $L(H)$, C_i, corresponding to the point u_i in H. Thus every point u_j, $j = 1, 2, ..., m$, determines a clique C_j of $L(H)$; it is clear that these cliques are all distinct.

We must now show that there are all the cliques of $L(H)$. Let C be any other clique of $L(H)$. If C has four or more points then H must contain four or more lines, any two of which must be adjacent, and the only way this can happen is for all four or more lines to have a point u_j in common. But this then would correspond to the clique C_j we have already constructed.

If C is a triangle then a similar argument shows that either H must contain a triangle, which contradicts our assumption that H has no triangles, or else the three lines of H which correspond to the three points of C all have a point u_k in common, which again would mean that $C = C_k$ is a clique we have already constructed.

The only remaining case to consider is that $L(H)$ contains a clique C having only one line, i.e., $C = K_2$. In this case H must contain two lines which are adjacent, say $u_i u_j$ and $u_j u_k$, where $u_i u_j$ and $u_j u_k$ correspond to the two points of C. But then $C = C_j$, the clique corresponding to the point u_j which we have already constructed.

We might pause for a moment to consider the trivial case that $L(H)$ contains a clique consisting of a single point. This would correspond in H to an isolated line. But since we are assuming that H is connected and contains at least three points, this situation cannot arise.

We now need to show that any two of the cliques C_1, C_2, ..., C_m of $L(H)$ are adjacent in $K(L(H))$ if and only if the two corresponding points from $u_1 u_2 \cdots u_m$ are adjacent. First suppose u_i and u_j are adjacent in H. Then it is clear that the point $u_i u_j$ in $L(H)$

belongs to both the clique C_i, which consists of all points of the form u_iw in $L(H)$, and the clique C_j, which consists of all points of the form u_jv. Thus cliques C_i and C_j are adjacent in $K(L(H))$, i.e., $C_i \cap C_j = \{u_iu_j\}$.

Conversely, suppose two cliques C_k and C_ℓ are adjacent in $K(L(H))$; then by our construction of C_k and C_ℓ, $C_k \cap C_\ell = \{u_ku_\ell\}$, i.e., there must exist a line in H connecting u_k and u_ℓ.

A number of corollaries now follow from Theorem 3, the first of which essentially rewords the theorem.

COROLLARY 3a. _If a connected graph_ G _has cliques_ C_1, C_2, $\ldots$, C_n _with_ $|C_i \cap C_j| \leq 1$ _for_ $1 \leq i, j \leq n$ _and_ $K(G)$ _has no triangles, then_ G _is isomorphic to the line graph_ $L(H)$ _where_ H _is the graph obtained from_ $K(G)$ _by adding points of degree one adjacent to the points_ C_i _in_ $K(G)$ _until for each_ i, _the new degree of point_ C_i _is made equal to_ $|C_i|$ _in_ G.

COROLLARY 3b. (Hedetniemi) _A graph_ G _is the line graph of a bipartite graph if and only if_ $|C_i \cap C_j| \leq 1$ _for any two cliques of_ G _and_ $K(G)$ _is a bipartite graph._

COROLLARY 3c. _A graph_ G _is the line graph of a tree if and only if_ $|C_i \cap C_j| \leq 1$ _for any two cliques of_ G _and_ $K(G)$ _is a tree._

COROLLARY 3d. _If_ T _is a tree having at least three points then_ $K^2(T) \cong T - \{u|\ \deg u = 1\}$.

COROLLARY 3e. _If_ B _is a bipartite graph then_ $K^2(B) \cong B - \{u|\ \deg u = 1\}$.

Let A be a $(0,1)$-matrix and let $H(A)$ be the row-column graph of A, i.e., a row R_i and a column C_j are adjacent if and only if the a_{ij} entry of A is one. It is easy to see that $H(A)$ is a bipartite graph; furthermore if every row and every column of A contain at least two ones, then $H(A)$ will have no points of degree 1.

COROLLARY 3f. (Cook) _If_ A _is a_ $(0,1)$-matrix _and_ $H(A)$ _is its row-column graph, then_ $K(L(H(A))) \cong H(A)$, _provided every row and every column contain at least two ones._

COROLLARY 3g. *If* T *is a* *tree* *having* *diameter* n, *then* $K^n(T) = K_1$.

COROLLARY 3h. *If a* *graph* G *contains a* *point* *which* *is* *adjacent* *to* *every* *other* *point* *of* G, *then* $K^2(G) = K_1$.

References

1. L. Beineke, Derived graphs and digraphs, *Beiträge zur Graphentheorie*, (H. Sachs, H. Voss, and H. Walther, eds.), Teubner, Leipzig, 1968, 17-33.

2. L. C. Chang, The uniqueness and nonuniqueness of the triangular association scheme, *Sci*. *Record* 3 (1959), 604-613.

3. C. Cook, Graphs Associated with (0,1) Arrays, Ph.D. Thesis, Univ. of Iowa, 1970.

4. F. Harary, *Graph* *Theory*, Addison-Wesley, Reading, Mass., 1969.

5. S. Hedetniemi, Graphs of (0,1)-matrices, *Recent* *Trends* *in* *Graph* *Theory*, (M. Capobianco, J. B. Frechen, and M. Krolik, eds.), Springer-Verlag, Berlin, 1971, 157-172.

6. A. J. Hoffman, On the uniqueness of the triangular association scheme, *Ann*. *Math*. *Statist*. 31 (1960), 492-497.

7. _____, On the line-graph of the complete bipartite graph, *Ann*. *Math*. *Statist*. 35 (1964), 883-885.

8. J. Krausz, Démonstration nouvelle d'une theoreme de Whitney sur les réseaux, *Mat*. *Fiz*. *Lapok* 50 (1943), 75-89.

9. J. Moon, On the line-graph of the complete bigraph, *Ann*. *Math*. *Statist*. 34 (1963), 664-667.

10. A. van Rooij and H. Wilf, The interchange graphs of a finite graph, *Acta* *Math*. *Acad*. *Sci*. *Hungar*. 16 (1965), 263-269.

11. N. Turri, Some observations on line-graphs, *Boll*. *Un*. *Math*. *Ital*. (3)20 (1965), 181-184.

LINE DIGRAPHS

Robert L. Hemminger
Vanderbilt University
Nashville, TN 37235

Whitney [11] introduced the concept of the line graph of a graph
in 1932. Since then the concept has received a great deal of study.
In 1960, Harary and Norman [7] extended this concept to line digraphs.
Several authors [1,2,3,4,7,8,9,10,12] have dealt with this concept. Almost all of the results for digraphs have to do with one of the following problems: (a) characterize the digraphs that are line digraphs,
(b) give a class of digraphs such that the line graph transformation
is a one-to-one function from this class onto the class of line digraphs, and (c) characterize those digraphs that are isomorphic to
their line digraphs. With the exception of Heuchenne [9] these results have been for finite multidigraphs (no loops allowed) or finite
digraphs (no loops or multiple edges allowed). The purpose of this
paper is to consider these problems for infinite pseudodigraphs. Satisfactory solutions have been given to the first two problems in the
finite case and their extension to the infinite case is not hard. The
third problem has not yet received a complete solution, even in the
finite case. Nor do we give a solution at this time. However, we do
anticipate a solution soon - at least in the finite case. Meanwhile
our results here do generalize the known finite results. In particular, we characterize the infinite functional and contrafunctional
pseudodigraphs and show that the weakly connected ones are isomorphic
to their line digraphs. We give some examples showing that the converse no longer holds in the infinite case and we give some other
results concerning this problem.

A _pseudodigraph_ D is a triple $(V(D),E(D),\Gamma)$ where $V(D)$ is a
nonempty set whose elements are called _vertices_, $E(D)$ is a set whose
elements are called _edges_, and Γ is a function from $E(D)$ into the
set of ordered pairs of vertices of D. If $\Gamma(\alpha) = \Gamma(\beta)$, α and β
are called _multiple edges_; if $\Gamma(\alpha) = (a,a)$, α is called a _loop_.
Rather than use the function Γ we will use the sets $S^+(a) =$
$\{\alpha \in E(D): \Gamma(\alpha) = (a,b)$ for some $b \in V(D)\}$ and $S^-(a) = \{\alpha \in E(D):$
$\Gamma(\alpha) = (b,a)$ for some $b \in V(D)\}$. Thus $S^+(a) \cap S^-(b)$ is the set of

multiple edges from a to b. The line digraph of a pseudodigraph D
with a nonempty edge set is denoted by L(D), has E(D) as its vertex set, and has $(\alpha, \beta) \in E(L(D))$ if and only if $\alpha = (a,b)$ and
$\beta = (b,c)$ for some $a,b,c \in V(D)$. Thus a line digraph has no multiple edges and has a loop at vertex α if and only if α is a loop
in D. We continue with the term line digraph even though it is
technically inconsistent (for a line digraph may now contain loops
while a digraph has no loops).

1. Characterizations of line digraphs. In this section we will
give some necessary and sufficient conditions for a pseudodigraph D
to be a line digraph. For this purpose, we define K(A,B), for
$A,B \subseteq V(D)$, to be the pseudodigraph with vertex set $A \cup B$ and with
one edge from each vertex of A to each vertex of B (note that we
do not require that $A \cap B = \phi$). A collection $\{S_i\}_{i \in I}$, of subsets
of a set S, some of which may be empty, is an improper partition of
S if $S = \bigcup_{i \in I} S_i$ and if distinct S_i are disjoint.

THEOREM 1.1. A pseudodigraph F is a line digraph of a pseudodigraph if and only if, for some set I, there exist two improper
partitions $\{A_i\}_{i \in I}$ and $\{B_i\}_{i \in I}$ of the vertices of F such that
$E(F) = \bigcup_{i \in I} (K(A_i, B_i))$.

Proof. Let D be a pseudodigraph. For each vertex a of D,
set $A_a = S^-(a)$ and $B_a = S^+(a)$. Then, in L(D), the subpseudodigraph induced by $A_a \cup B_a$ is $K(A_a, B_a)$, and if (α, β) is an edge in
L(D) then $\alpha = (a,b)$ and $\beta = (b,d)$ in D, so (α, β) is an edge
in $K(A_b, B_b)$. The necessity of the condition follows by taking
$I = V(D)$.

Conversely, let F be a pseudodigraph satisfying the condition
of the theorem. Let D be the pseudodigraph with the ordered pairs
(A_i, B_i) as vertices and with $|B_i \cap A_j|$ edges from (A_i, B_i) to
(A_j, B_j) for each i and j (including i = j). Let $\sigma_{i,j}$ be a
one-to-one function from $B_i \cap A_j$ onto this set of edges of D. Then
the function σ defined on V(F) by taking σ to be $\sigma_{i,j}$ on
$B_i \cap A_j$ is a well-defined function of V(F) into V(L(D)), since
$\{B_i \cap A_j\}_{i,j \in I}$ partitions V(F). Moreover σ is one-to-one and onto
since each $\sigma_{i,j}$ is, and one easily sees that σ is an isomorphism
from F onto L(D); e.g. if $(a,b) \in E(F)$, then there are i, j,

and k such that $a \in B_i \cap A_j$ and $b \in B_j \cap A_k$. Thus $\sigma(a)$ is an edge of D from (A_i,B_i) to (A_j,B_j) and $\sigma(b)$ is an edge of D from (A_j,B_j) to (A_k,B_k). Hence $(\sigma(a),\sigma(b)) \in E(L(D))$.

Thus we have, as corollaries, the following results which appear in [1,7].

COROLLARY 1.2. A pseudodigraph F is the line digraph of a multidigraph (resp., a pseudodigraph without multiple edges; resp., a digraph) if and only if the condition of Theorem 1.1 is satisfied and if in addition, for each i, $B_i \cap A_i = \phi$ (resp., for each i and j, $B_i \cap A_j$ has at most one element; resp., for each i and j, $B_i \cap A_j$ has at most one element and has none if i = j).

A more useful result is given by Heuchenne [9] (also, see [7],[10]).

THEOREM 1.3. A pseudodigraph D without multiple edges is a line digraph of a pseudodigraph if and only if whenever (a,b), (c,b), and (c,d) are edges of D, then so is (a,d).

The vertices a, b, c, and d in Theorem 1.3 need not be distinct. The non-trivial ways that this situation can occur for a pseudodigraph without multiple edges are given in Figure 1, where the existence of the bold edges implies the existence of the dotted edge.

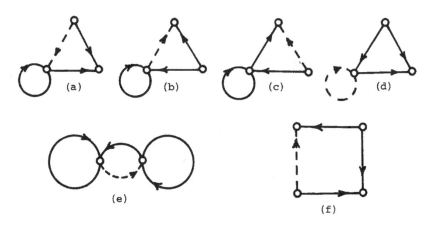

Figure 1.

There are five more edges that could be added to the pseudodi-graph in (a) of Figure 1 without the dotted edge and without intro-ducing multiple edges. Thus there are $2^5 = 32$ pseudodigraphs with three vertices and without multiple edges that contain the pseudodi-graph in (a) (but not the dotted edge) that are not induced subpseudo-digraphs of a line digraph. Similarly, there are 32 forbidden induced subpseudodigraphs associated with each of those in (b), (c), (d) (however some of those arising from (a), (b), (c), and (d) have been counted twice), and $2^{12} = 4096$ arising from (f). We thus have

THEOREM 1.4. _A pseudodigraph_ D _without multiple edges is a line digraph of a pseudodigraph if and only if none of the pseudodi-graphs associated with the six pseudodigraphs of Figure 1 is an in-duced subpseudodigraph of_ D.

From this we easily get two results given by Beineke [2].

COROLLARY 1.5. _A pseudodigraph_ D _without multiple edges is the line digraph of a digraph if and only if, in addition to the forbidden pseudodigraphs in Theorem 1.4, the first three digraphs of Figure 2 are also forbidden as induced subdigraphs of_ D.

The first of the digraphs in Figure 2 excludes loops and the next two exclude multiple edges.
An _oriented graph_ is a digraph with at most one edge (in either direction) between a pair of vertices. One easily sees that a pseudo-digraph D is the line digraph of an oriented graph if and only if D is the oriented line digraph of a pseudodigraph, so we have

COROLLARY 1.6. _A pseudodigraph_ D _without multiple edges is the line digraph of an oriented graph if and only if none of the digraphs of Figure 2 is an induced subdigraph of_ D.

Let H_n be the (n-1)-st subdivision of one of the pseudodigraphs of Figure 1, i.e., each (dotted) edge of the pseudodigraph is replaced by a (dotted) dipath of length n. Then we will say that a pseudodi-graph D satisfies that n^{th}-Huchenne condition if and only if the existence of the bold edges of H_n in D, for any such H_n, implies the existence of the dotted dipath of H_n in D (note that the dotted dipath is openly disjoint from the rest of H_n). The following generalization of Huchenne's Theorem is of interest in studying iterated line digraphs, i.e., $L^n(D)$ where $L^1(D) = L(D)$ and $L^n(D) = L(L^{n-1}(D))$ for $n > 1$.

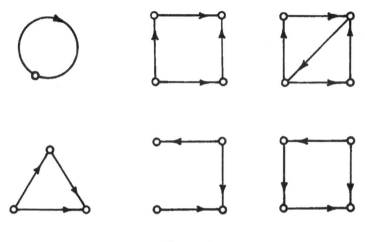

Figure 2.

THEOREM 1.7. Let F be a pseudodigraph without multiple edges. Then $F = L^n(D)$ for some pseudodigraph D if and only if F satisfies the first n Heuchenne conditions.

The proof is a straightforward induction argument.

2. Realization of line digraphs. As pointed out in [5] the two major problems to be considered for a graph transformation are characterization and realization. In this section we treat the second of these problems.

For a pseudodigraph D let $\tilde{D}$ be the subpseudodigraph induced by the vertices of D that are neither sinks nor sources.

THEOREM 2.1. Let D and F be a pseudodigraphs and let σ be an isomorphism of L(D) onto L(F). Then σ restricted to $E(\tilde{D})$ is induced, in the natural way, by an isomorphism of $\tilde{D}$ onto $\tilde{F}$. Also α is a source (resp., sink) edge of D if and only if $\sigma(\alpha)$ is a source (resp., sink) edge of F.

Proof. If $a \in V(\widetilde{D})$, say $\alpha \in S^-(a)$ and $\beta \in S^+(a)$, then there exists an $a' \in V(\widetilde{F})$ with $\sigma(\alpha) \in S^-(a')$ and $\sigma(\beta) \in S^+(a')$. If $\gamma \in S^-(a)$ then $(\sigma(\gamma), \sigma(\beta))$ is an edge of F, so $\sigma(\gamma) \in S^-(a')$, i.e., $\sigma(S^-(a)) \subseteq S^-(a')$. Similarly, $\sigma(S^+(a)) \subseteq S^+(a')$. Since $\sigma^{-1}(\sigma(\alpha)) \in S^-(a)$ and $\sigma^{-1}(\sigma(\beta)) \in S^+(a)$, and since σ^{-1} is an isomorphism of $L(F)$ onto $L(D)$, we have, in the same manner, $\sigma^{-1}(S^-(a')) \subseteq S^-(a)$ and $\sigma^{-1}(S^+(a')) \subseteq S^+(a)$. It follows that $\sigma(S^-(a)) = S^-(a')$ and $\sigma(S^+(a)) = S^+(a')$.

Since, for $a,b \in V(\widetilde{D})$ (or $V(\widetilde{F})$), $S^+(a) = S^+(b)$ if and only if $a = b$, the function σ^* defined by the equation $\sigma(S^+(a)) = S^+(\sigma^*(a))$ is a well-defined one-to-one function of $V(\widetilde{D})$ into $V(\widetilde{F})$. And σ^* is onto $V(\widetilde{F})$; if $b' \in V(\widetilde{F})$ then, by symmetry, $\sigma^{-1}(S^+(b')) = S^+(a)$ for some $a \in V(\widetilde{G})$ and so $b' = \sigma^*(a)$. Moreover, for $a,b \in V(\widetilde{D})$ ($a = b$ allowed) we have $\sigma(S^+(a) \cap S^-(b)) = S^+(\sigma^*(a)) \cap S^-(\sigma^*(b))$; hence $|S^+(a) \cap S^-(b)| = |S^+(\sigma^*(a)) \cap S^-(\sigma^*(b))|$, so σ^* is an isomorphism.

The last statement of the theorem now follows immediately.

Note that the theorem is still valid if $V(\widetilde{D}) = \phi$. The possibilities with $V(\widetilde{D}) = \phi$ emphasize how very different D and F can be and still have $L(D) \simeq L(F)$.

The above theorem was given for finite multidigraphs by Harary and Norman [7]). It has as an immediate consequence an unpublished result of Zelinka that generalizes a finite digraph theorem of Aigner [1].

COROLLARY 2.2. Let P be the class of connected pseudodigraphs and let Q be the class of connected pseudodigraphs having at most one source and at most one sink. Then

(a) for each $D \in P$, there is an $F \in Q$ such that $L(D) \simeq L(F)$, and

(b) if F_1 and $F_2 \in Q$, then $F_1 \simeq F_2$ if and only if $L(F_1) \simeq L(F_2)$.

3. Digraphs isomorphic to their line digraphs. A pseudodigraph D is functional if $|S^+(a)| = 1$ for each $a \in V(D)$; it is contrafunctional if its converse, which we denote by D^r, is functional, i.e., if $|S^-(a)| = 1$ for each $a \in V(D)$. Harary and Norman [7] showed that line digraphs are intimately tied up with functional and contrafunctional digraphs by showing that, for finite weakly connected

digraphs, $D \sim L(D)$ if and only if D or D^r is functional. We shall see later that this result is not valid for infinite digraphs; however infinite functional digraphs are still of interest in studying pseudodigraphs that are isomorphic to their line digraph, so we give a characterization of them in this section. But first we give a few preliminary results.

LEMMA 3.1. For pseudodigraphs D and F, $D \sim F$ if and only if $D^r \sim F^r$.

Proof. Let σ be a one-to-one function of $V(D)$ onto $V(F)$. Then $|S^+(a) \cap S^-(b)| = |S^+(\sigma(a)) \cap S^-(\sigma(b))|$ for all $a,b \in V(D)$ if and only if $|S^-(a) \cap S^+(b)| = |S^-(\sigma(a)) \cap S^+(\sigma(b))|$ for all $a,b \in V(D^r)$. The lemma follows.

LEMMA 3.2. For a pseudodigraph D we have $L(D^r) \sim L(D)^r$.

Proof. Note that $\alpha \in V(L(D))$ if and only if $\alpha \in E(D)$, i.e. $\alpha^r \in E(D^r)$, i.e. $\alpha^r \in V(L(D^r))$. Hence the function $\sigma: V(L(D)) \to V(L(D^r))$ defined by $\sigma(\alpha) = \alpha^r$ is one-to-one and onto. The following equivalent statements show that σ is an isomorphism: $(\alpha,\beta) \in E(L(D^r))$; $\alpha = (a,b)$, $\beta = (b,c) \in E(D^r)$ for some $a,b,c \in V(D^r)$; $\alpha^r = (b,a)$, $\beta^r = (c,b) \in E(D)$ for some $a,b,c \in V(D)$; $(\beta^r,\alpha^r) \in E(L(D))$; $(\alpha^r,\beta^r) \in E(L(D)^r)$.

COROLLARY 3.3. For pseudodigraphs D and F we have $F \sim L(D)$ if and only if $F^r \sim L(D^r)$, i.e. F is a line digraph if and only if F^r is.

Let D be a weakly connected functional pseudodigraph. Thus D has no sinks. Let $(a_0,a_1) \in E(D)$. If $a_1 \neq a_0$, pick $a_2 \in V(D)$ (if possible) so that (a_0,a_1,a_2) is a dipath in D. If this procedure is blocked it is because, for some k, $(a_k,a_{k+1},\ldots,a_k)$ is a dicycle in D (we include a loop as a dicycle - of length one); otherwise $(a_0,a_1,a_2,\ldots)$ is a one-way infinite dipath in D. In the latter case extend this dipath backward from a_0 as far as possible. If this procedure is not blocked, then D has a 2-way infinite dipath (we will refer to these as dicycles of infinite length); otherwise D has a one-way infinite dipath $(a_0,a_1,a_2,\ldots)$ where a_0 is a source of D.

Suppose that $C = (a_0, a_1, \ldots, a_n, a_0)$ is a dicycle in D. Let D' be the subdigraph of D consisting of C and all vertices and edges of D that are on dipaths from vertices of D to vertices of C. Since D is functional there is a unique dipath from a vertex of D to C. Thus D' consists of C plus disjoint arborescences rooted at vertices of C (an _arborescence_ A rooted at a is a tree oriented so that there is a dipath to a from each vertex of A; a _counterarborescence_ rooted at a is the converse of an arborescence rooted at a). If $D \neq D'$ then, since D is weakly connected, there is an edge (a,b) with $a \in V(D')$, $b \notin V(D')$. But this contradicts the fact that $|s^+(a)| = 1$, so $D = D'$. Treating the other two classes in a similar fashion we have

THEOREM 3.4. Let D _be a weakly connected pseudodigraph. Then_ D _is functional if and only if_ D _is the union of a subdigraph_ C _and disjoint arborescences rooted at vertices of_ C _where_ C _is either_ (a) _a finite dicycle,_ (b) _an infinite dicycle, or_ (c) _a one-way infinite dipath from a source of_ D.

The converse is obvious.

Harary's [6] characterization of finite weakly connected functional digraphs follows immediately. The following result is also an easy consequence of Lemma 3.2 and Theorem 3.4.

COROLLARY 3.5. _If_ D _is a functional or contrafunctional pseudo-digraph, then_ $L(D) \simeq D$.

Several authors [1, 2, 3, 4, 7, 9,12] have dealt with the problem of characterizing finite digraphs D with $D \simeq L(D)$, but Harary and Norman [7] were the first to show that, for finite weakly connected digraphs, the converse of Corollary 3.5 is valid. In the remainder of the paper we will discuss, but not characterize, pseudodigraphs D with $D \simeq L(D)$.

First we note that the digraph in Figure 3 shows that the converse of Corollary 3.5 is not valid for infinite weakly connected digraphs. This example also shows that Beineke's [2] statement about unicyclic digraphs of _period_ one (i.e. for some n, $L^n(D) \simeq L^{n+1}(D)$) is not true for infinite digraphs; in fact, if one deletes all vertices of this digraph at a distance greater than one from the dicycle, the resulting digraph is a counterexample to the statement.

The following result is due to Heuchenne [9].

157

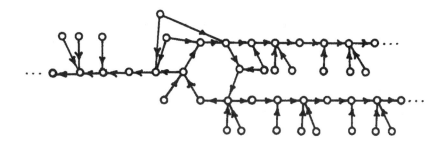

Figure 3.

LEMMA 3.6. If D is a weakly connected pseudodigraph with D ≃ L(D), then D does not have both sources and sinks.

Proof. If L(D) has both sources and sinks then it has a pair, say source α and sink β, that are as close together as any other pair. Let $[\alpha = \alpha_0, \alpha_1, \alpha_2, \ldots, \alpha_n = \beta]$ be a shortest path (not necessarily a dipath) between α and β in L(D). If this path is a dipath, then $((\alpha_0, \alpha_1), (\alpha_1, \alpha_2), \ldots, (\alpha_{n-1}, \alpha_n))$ is a source to sink dipath in $L^2(D) \simeq L(D)$ of length less than n, which is a contradiction. So suppose the path is not a dipath. Since no α_i, $i = 1, 2, \ldots, n-1$ is a source or sink in L(D), $[\alpha_0, \alpha_1, \ldots, \alpha_n]$ is the edge sequence of a semiwalk in D; and if each edge α_i $i = 1, 2, \ldots, n$, with (α_{i-1}, α_i) and $(\alpha_{i+1}, \alpha_i) \in E(L(D))$, or with (α_i, α_{i-1}) and $(\alpha_i, \alpha_{i+1}) \in E(L(D))$, is deleted, the result is a source to sink path in D. But there are at least two such change of direction edges, so we have a source to sink path in D ≃ L(D) of length less than n. Once again this is a contradiction, so D does not have both sources and sinks.

If D has no sinks the function σ: E(D) → V(D) given by σ(a,b) = a is onto V(D). If D is finite and D ≃ L(D), then σ is one-to-one since $|E(D)| = |V(L(D))| = |V(D)|$. Thus, by Corollary 3.5 we have the Harary-Norman [7] result for finite digraphs and the Heuchenne [9] result for finite pseudodigraphs.

THEOREM 3.7. Let D be a finite weakly connected pseudodigraph. Then D ≃ L(D) if and only if D or D^r is functional.

We see, by Corollary 3.5 and part (c) of Theorem 3.4, that, in contrast to the finite case, digraphs without dicycles (finite or infinite) can be isomorphic to their line digraphs. However, we will see in Theorem 3.10 that these are still either functional or contra-functional.

LEMMA 3.8. If D is weakly connected and $D \simeq L(D)$, then D has at most one dicycle of finite length.

Proof. Suppose D has two finite dicycles with vertices in common. Let n be the smallest positive integer such that D has two finite dicycles, with vertices in common, that have n as the sum of their lengths. Let m be the maximum number of vertices that two dicycles, with the sum of their lengths equal to n, have in common. Let C_1 and C_2 be two such dicycles. If $C_1 = (a_1, a_2, \ldots, a_p, a_1)$ has a diagonal (a_i, a_j), then, by the first-Heuchenne condition (a_{j-1}, a_{i+1}) is an edge of D and hence one of the dicycles $C_1' = (a_i, a_j, a_{j+1}, \ldots, a_{i-1}, a_i)$ or $C_1'' = (a_{j-1}, a_{i+1}, a_{i+2}, \ldots, a_{j-2}, a_{j-1})$ contains vertices in common with C_2. But this contra-dicts the choice of n. Thus neither C_1 nor C_2 have diagonals. But, for C_1 and C_2 in $L(D)$, this means that there are dicycles C_i' in D such that $L(C_i') = C_i$, i = 1, 2. (Note that there is al-ways such a closed ditrail in D; but it is a dicycle if and only if C_i has no diagonals in $L(D)$.) But C_i' has the same length as C_i. Moreover C_1' and C_2' have more vertices in common than C_1 and C_2 do, since they have the same number of edges in common as C_1 and C_2 have vertices in common. This contradicts the choice of m. Hence every pair of finite dicycles in D are disjoint.

But, if $[\alpha_0, \alpha_1, \ldots, \alpha_n]$ is a path (not necessarily a dipath) in $L(D)$ between two disjoint dicycles, then $[\alpha_1, \alpha_2, \ldots, \alpha_{n-1}]$ is the edge sequence of a semiwalk between two dicycles in D; i.e., if there are dicycles in $D \simeq L(D)$ at distance $n \geq 1$ apart then there are dicycles at distance less than n apart. It follows that there are distinct dicycles with vertices in common if there are distinct dicycles in D. In light of the last paragraph, the lemma now follows.

COROLLARY 3.9. If D is weakly connected with $D \simeq L(D)$ and if D has both finite and infinite dicycles, then D has an infinite number of infinite dicycles.

Proof. By Lemma 3.8, D has a unique finite dicycle C. One easily sees that there are dicycles disjoint from C. But if C' is one at distance n from C, then L(C') is at a distance greater than n from L(C), the unique finite dicycle.

THEOREM 3.10. **If** D **is a weakly connected pseudodigraph with** $D \sim L(D)$ **and without dicycles, then** D **or** D^r **is functional.**

Proof. Without loss of generality, we can assume, by Lemma 3.6, that D has no sinks. Then, as in the first paragraph of the proof of Theorem 3.4, D has a one-way infinite dipath $P = (a_0, a_1, \ldots)$ from a source of D.

Let T_i be the subdigraph of D with edges and vertices being those contained in dipaths from vertices of D to a_i, $i \geq 1$, that do not contain either a_{i-1} or a_{i+1}.

Suppose that L(D) contained vertices a and b with two distinct dipaths from a to b. Then we can assume that these dipaths are in fact openly disjoint. But the preimage, under L, of this configuration contains a smaller such configuration. Since L(D) does not contain multiple edges, repetition of this procedure leads us to conclude that L(D) has an edge (a,b) and a dipath (a, b_1, b_2, ..., b_n, b) with $n \geq 1$. But then, by the first-Heuchenne condition, $(b_1, b_2, \ldots, b_n, b_1)$ is a dicycle in D. However, D has no dicycles; so, for each a,b $\in$ V(D), there is at most one dipath in D from a to b.

It follows from the last paragraph that T_i is an arborescence rooted at a_i and that T_i and T_j are disjoint if $i \neq j$.

Suppose D has a vertex of out-degree greater than one. Then there is such a vertex in P or in some T_i. By use of L^{-1} it follows that there is such a vertex in P. Thus there is a smallest positive integer n such that there is a dipath of length n, from a source to a vertex with out-degree greater than one. But the image of this dipath, under L, is a dipath of length n - 1 from a source to a vertex with out-degree greater than one. This contradicts the fact that $D \sim L(D)$. The theorem follows, since D is just P plus the T_i. We note that, since D has no infinite dicycles, the T_i's are finite.

If $D \sim L(D)$ and D has two infinite dicycles with only a finite number of vertices in common, then one easily sees that D has an infinite number of pairwise disjoint infinite dicycles at arbitrarily large distances apart. Such an example is given in Figure 6.

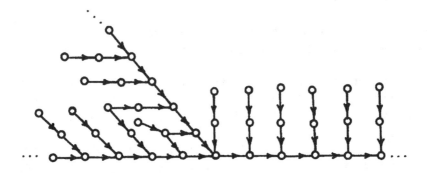

Figure 4.

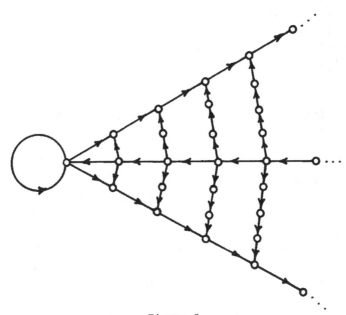

Figure 5.

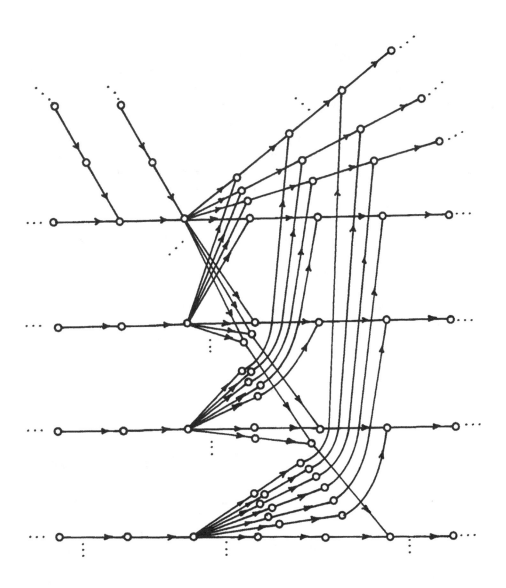

Figure 6.

Thus, if D only has a finite number of infinite dicycles (hence D has no finite dicycles by Corollary 3.9), then they all have their right "ends" in common or else they all have their left "ends" in common. As before we have at most one dipath between two vertices of D, since we have no finite dicycles. Thus the rest of D consists of finite arborescences or counterarborescences at vertices of the infinite dicycles. However, by Lemma 3.6, not both arborescences and counterarborescences occur. One easily sees that all of these arborescences (counterarborescences) must be isomorphic. Also portions of these arborescences (counterarborescences) must be identified so that the nth-Heuchenne condition is satisfied. The identifications to be made depend on distances to a vertex where two of the infinite dicycles come together. An example of this type of a digraph is given in Figure 4. The example illustrates that digraphs of this type need not be functional or contrafunctional.

There are three types of weakly connected infinite pseudodigraph with $D \sim L(D)$ not characterized in this paper: (1) those with a finite dicycle and no infinite dicycles, (2) those with a finite dicycle and an infinite number of infinite dicycles, and (3) those with no finite dicycles and an infinite number of infinite dicycles.

Figure 3 contains an example of type (1), Figure 5 contains an example of type (2), and Figure 6 contains an example of type (3)

References

1. A. Aigner, On the line-graph of a directed graph, _Math. Z._ 102 (1967), 56-61, MR36, #76.

2. L. Beineke, On derived graphs and digraphs, _Beitrage zur Graphentheorie_ (Ed. H. Sachs, H. J. Voss, H. Walther), Teubner, Leipzig 1968, pp. 17-23.

3. A. Ghirlanda and L. Muracchini, Sui grafi segnati ed i grafi commutati, _Statistica_ (Bologna) 25 (1965), 677-680, MR33, #7272.

4. _____, Sul grafo commutato e sul grato opposto di un grafo orientato, _Atti Sem. Mat. Fis. Univ. Modena_ 14 (1965), 87-97, MR33, #2570.

5. B. Grünbaum, Incidence patterns of graphs and complexes, _The Many Facets of Graph Theory_, (G. Chartrand and S. F. Kapoor, ed.), Springer-Verlag, Berlin, 1969, 115-128, MR40, #4152.

6. F. Harary, The number of functional digraphs, _Math. Analen_ 138 (1959), 203-210, MR22, #18.

7. F. Harary and R. Norman, Some properties of line digraphs, _Rend._
    Circ. _Mat_. _Palermo_ 9 (1960), 161–168, MR24, #A693.

8. F. Harary and D. Geller, Arrow diagrams are line digraphs, _SIAM_
    _J_. _Appl_. _Math_. 16 (1968), 1141–1145, MR38, #3179.

9. C. Heuchenne, Sur une certaine correspondance entre graphes,
    Bull. _Soc_. _Roy_. _Sci_. _Liege_ 33 (1964), 743–753, MR30, #5297.

10. P. Richards, Precedence constraints and arrow diagrams, _SIAM_
    Review 9 (1967), 548–553.

11. H. Whitney, Congruent graphs and the connectivity of graphs,
    Amer. _J_. _Math_. 54 (1932), 150–168.

12. C. Zamfirescu, Disconnected digraphs isomorphic with their line
    digraphs, _Bollettino_ _U.M.I_. (4) 4 (1971), 888–894.

ON LIMIT POINTS OF SPECTRAL RADII
OF NON-NEGATIVE SYMMETRIC INTEGRAL MATRICES*

Alan J. Hoffman
Mathematical Sciences Department
IBM Watson Research Center
Yorktown Heights, NY

1. **Introduction.** Let G be the set of all symmetric matrices of all orders, every entry of which is a non-negative integer. Each such matrix A has a largest eigenvalue, denoted by $\rho(A)$, which is positive unless $A = 0$. Let $R = \{\rho \mid \rho = \rho(A)$ for some $A \in G\}$. We pose the problem of finding all limit points of R, and offer the following contribution.

THEOREM. Let $\tau = (\sqrt{5} + 1)/2$ (the golden mean). For $n = 1, 2, \ldots,$ let β_n be the positive root of

$$P_n(x) = x^{n+1} - (1 + x + x^2 + \cdots + x^{n-1}).$$

Let $\alpha_n = \beta_n^{1/2} + \beta_n^{-1/2}$. Then

$$2 = \alpha_1 < \alpha_2 < \cdots$$

are all limit points of R smaller than $\tau^{1/2} + \tau^{-1/2} = \lim_n \alpha_n$.

In Section 2, we show that it is sufficient to restrict G to the set of (0,1) symmetric matrices with 0 diagonal, i.e., to the adjacency matrices of graphs. Then we show in Section 3 that if $\rho < \tau^{1/2} + \tau^{-1/2}$ is a limit point of R, we need only consider graphs which are trees. In Section 4, we show we need only consider trees in which there is at most one vertex of degree three (and no vertex of higher degree), and this vertex is adjacent to an end vertrex of the tree.

In Section 5, we complete the algebra needed to establish the formula given in the theorem, and we state some unsolved problems in Section 6.

---

*This work was supported (in part) by the Army Research Office under contract number DAHC04-72-C-0023.

This manuscript has benefited from help by T. J. Rivlin on polynomials and recurrences, especially in Section 5, and from John Smith, who simplified and generalized the content of Section 4.

2. Reduction to graphs.

PROPOSITION 2.1. Let $B \in G$. Then there exists a graph G such that $\rho(A(G)) = \rho(B)$.

Proof. It is sufficient to assume B irreducible. Let $\rho(B) = \rho$, $Bx = \rho x$, $x > 0$. Let $B_1, B_2, \ldots, B_k$ be symmetric $(0,1)$ matrices such that $\Sigma B_i = B$. Let

$$C = \begin{bmatrix} B_1 & B_2 & \cdots & B_k \\ B_2 & B_3 & \cdots & B_1 \\ & & \cdots & \\ B_k & B_1 & \cdots & B_{k-1} \end{bmatrix}$$

and $A = \begin{pmatrix} 0 & C \\ C & 0 \end{pmatrix}$. Then A is a symmetric $(0,1)$ matrix with 0 diagonal; hence, A is the adjacency matrix of a graph. Let $y = (x, x, \ldots, x)$. From $B = \Sigma B_i$, it follows that $Ay = \rho y$. Since $y > 0$, it follows that $\rho = \rho(A)$.

3. Reduction to trees. Henceforth, for any graph G, we will write $\rho(G)$ for $\rho(A(G))$. If G and H are graphs, we write $G \prec H$ (i.e., G is a subgraph of H) if every vertex of G is a vertex of H and every edge of G is an edge of H. The following two lemmas come from the Perron-Frobenius theory of non-negative matrices.

LEMMA 3.1. If $G \prec H$, $\rho(G) \leq \rho(G)$. If $G \prec H$, $G \neq H$, with H connected, then $\rho(G) < \rho(H)$.

LEMMA 3.2. If G is connected, with $x > 0$, then $A(G)x \leq \mu x$; if $A(G) \neq \mu x$, then $\mu > \rho(G)$.

LEMMA 3.3. If G connected, σ is an automorphism of G, and $A(G)x = \rho x$, then $\sigma i = j$ implies $x_i = x_j$.

Proof. Let S be the permutation matrix corresponding to σ, $A = A(G)$. Then $\rho x = Ax = S^T A S x$, $x > 0$. Therefore, $\rho S x = A S x$; hence, Sx is a Perron vector corresponding to A. But since Sx is a positive vector with the same norm as x, the essential uniqueness of the Perron vector shows that $Sx = x$.

PROPOSITION 3.4. <u>The smallest limit point in</u> R <u>is</u> 2.

Proof. Let G_1, G_2, ... be a sequence of graphs such that $\rho(G_i) \neq \rho(G_j)$ for $i \neq j$, and $\rho(G_n) \rightarrow \lambda < 2$. Suppose G is a connected graph on at least three vertices, the maximum degree of the vertices of G is $\Delta(G)$, and the diameter of G is $d(G)$. Then $|V(G)| \leq \Delta^d + 1$. Therefore, $\max(\Delta(G), d(G)) \geq (\log|V(G)| - 1)^{1/2}$. But since the graphs G_i are all different, $|V(G_i)| \rightarrow \infty$. Hence, for sufficiently large n, G_n contains as a subgraph an arbitrarily long path S_k or an arbitrarily large claw $K_{1,t}$. But $\rho(K_{1,t}) \rightarrow \infty$ and $\rho(S_k) \rightarrow 2$.

LEMMA 3.4. <u>Let</u> A_{-1} <u>be a principal submatrix of order</u> $n - 1$ <u>of a symmetric matrix</u> A_0 <u>of order</u> n <u>with non-negative entries.</u> <u>Define</u> A_i <u>recursively by</u>

$$
A_i = \begin{bmatrix} & & & & 0 \\ & & & & 0 \\ & A_{i-1} & & & \cdot \\ & & & & \cdot \\ & & & & 0 \\ & & & & 1 \\ 0 & 0 & \ldots & 0 & 1 & 0 \end{bmatrix}.
$$

<u>Assume further that</u> $\lim\limits_{i \to \infty} \rho(A_i) > 2$. <u>Then</u> $\lim\limits_{i \to \infty} \rho(A_i) = $ <u>largest positive root of</u> $\theta P - Q$, <u>where</u> $\theta = (x + \sqrt{x^2-4})/2$, P <u>is the characteristic polynomial of</u> A_0 <u>and</u> Q <u>is the characteristic polynomial of</u> A_{-1}.

Proof. If P_i is the characteristic polynomial of A_i, we have $P_i = xP_{i-1} - P_{i-2}$. Further, $\rho(A_i)$ is monotonically increasing by Lemma 3.1, and is bounded since $\rho(A_i)$ is at most the maximum row sum of A_i (by Lemma 3.2), which is bounded. Therefore, $\rho(A_i)$

exists. The rest follows from the theory of linear difference equations.

PROPOSITION 3.6. <u>Let</u> T_n <u>be the tree</u>

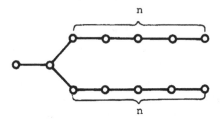

<u>Then</u> $\lim_{n} \rho(T_n) = \tau^{1/2} + \tau^{-1/2}$.

<u>Proof</u>. Since $\rho(T_3) = 2$, it follows from Lemma 3.1 that $\lim_{n} \rho(T_n) > 2$. Using Lemma 3.3, we can apply Proposition 3.5 with

$$A_0 = \begin{bmatrix} 0 & 1 & 0 \\ 1 & 0 & 2 \\ 0 & 1 & 0 \end{bmatrix}, \qquad A_{-1} = \begin{bmatrix} 0 & 1 \\ 1 & 0 \end{bmatrix}.$$

(Note that A_0 and all A_i are similar to symmetric matrices, so Proposition 3.5 applies.) Then the biggest root of $\theta P - Q$ is $\sqrt{2} + \sqrt{5} = \tau^{1/2} + \tau^{-1/2}$.

PROPOSITION 3.7. <u>If</u> G <u>is a connected graph which is neither a tree nor a circuit, then</u> $\rho(G) > \tau^{1/2} + \tau^{-1/2}$.

<u>Proof</u>. By hypothesis, G contains

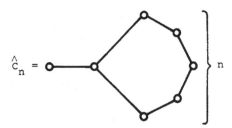

as a proper subgraph. We will prove $\rho(\hat{C}_n) > \rho(\hat{C}_{n+1})$. Assume this has been done. Since, by Lemma 3.1, $\rho(G) \geq \rho(\hat{C}_n) > \rho(T_{[n/2]})$, we would have $\rho(G) \geq \rho(\hat{C}_n) > \lim_n \rho(\hat{C}_n) \geq \lim_n \rho(T_n) = \tau^{1/2} + \tau^{-1/2}$ by Proposition 3.6.

There are two cases in proving $\rho(\hat{C}_n) > \rho(\hat{C}_{n+1})$, depending on whether n is odd or even. To illustrate, we do $n = 4$. Consider $\hat{C}_4$, where the labels at the vertices refer to the coordinate of the Perron vector for $\hat{C}_4$. Next, put the following coordinates at $\hat{C}_5$ (see Lemma 3.3).

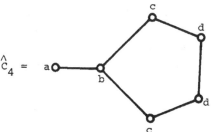

$$\hat{C}_4 =$$

Since $\rho(\hat{C}_4) > 2$, it will be seen that this vector plays the role of x in Lemma 3.2 applied to $A(\hat{C}_5)$, and $\rho(\hat{C}_4)$ plays the role of μ. A similar argument applies if n is odd.

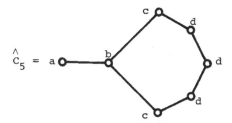

$$\hat{C}_5 =$$

4. __Restrictions on trees__. For the material in this section, we need two results, which we shall prove elsewhere.

Let e be an edge of a graph G. If there exists a path in G, $x_1, x_2, \ldots, x_k$ where x_{k-1} and x_k are the end vertices of e, and where the degrees of $x_1, \ldots, x_{k-1}$ are respectively 1, 2, 2, $\ldots$, 2, then e is said to be on an __end path__ of G.

PROPOSITION 4.1. __Let__ G __be a connected graph with__ $\rho(G) > 2$, e __an edge of__ G __(with vertices__ x, y) __not on an end path of__ G, G __not a simple polygon. Let__ $G_{x,y}$ __be the graph obtained from__ G __by__

deleting edge e, and adding a vertex z adjacent to x and y,
but to no other vertices of G. Then $\rho(G_{x,y}) < \rho(G)$.

Next, let G be a connected graph, v a vertex of G of
degree at least two, and let (G,v,n) be the graph obtained from G
by appending a path of n vertices to G at v. Define $\rho(G,v) =$
$\lim_{n\to\infty} \rho(G,v,n)$. If P(G) is the characteristic polynomial of A(G),
then Lemma 3.5 asserts that $\rho(G,v) =$ largest root of $\theta P(G) - P(G-v)$,
where $\theta = (x + \sqrt{x^2-4})/2$.

Let G_1, G_2 be disjoint connected graphs, v_1 a vertex of
degree at least two of G_1, v_2 a vertex of degree at least two
of G_2, and (G_1, v_1, n, v_2, G) the graph obtained from G_1 and
G_2 by joining them by a path of n vertices connecting v_1 and v_2.
Define $\rho(G_1, v_1, G_2) = \lim_{n\to\infty} \rho(G_1, v_1, n, v_2, G_2)$.

PROPOSITION 4.2. Given G_1, v_1, v_2, G_2 as above, $\rho(G_1, v_1,$
$v_2, G_2) = \max(\rho(G_1, v_1), \rho(G_2, v_2))$.

Assume $G_1, G_2, G_3, \ldots$ a sequence of distinct trees such that
$\rho(G_i) \to \lambda$, where $2 < \lambda < \tau^{1/2} + \tau^{-1/2}$. Assume an infinite number
have vertices of degree at least 4. Then

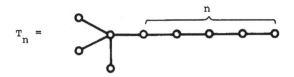

is for each n a partial subtree of G_i for an infinite number of
i. But $\rho(T_n) \to 4/\sqrt{3} > \tau^{1/2} + \tau^{-1/2}$, a contradiction.

So we may assume (by going to subsequences) that the maximum
degree of each G_i is at most 3. Suppose some G_i has at least
three vertices of degree three. Then

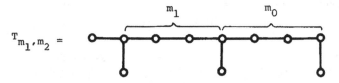

is a subtree for some values of m_1 and m_2. By Proposition 4.1, $\rho(T_{m_1,m_2}) > \rho(T_{n,n})$ for arbitrarily large n. By Proposition 3.6, this implies $\rho(G_i) > \tau^{1/2} + \tau^{-1/2}$, a contradiction.

So we may assume that each G_i contains at most two vertices of degree three. Assume an infinite number contain two such vertices, but the distance between them is bounded. This means that, for some h,

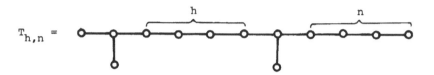

$$T_{h,n} =$$

is a subtree for arbitrarily large n. By Proposition 4.1, $\rho(T_{h,n}) > \rho(T_n)$, where T_n is the graph of Proposition 3.6, a contradiction.

Therefore, the distance between the two vertices of degree 3 is unbounded. By Propositions 3.6 and 4.2, we could ignore one of the vertices of degree 3.

Hence, we may assume that each G_i is a tree with exactly one vertex of degree three. Suppose two or more of the end paths become arbitrarily long. Again, we violate Proposition 3.6. Suppose two of the end paths contain at least two vertices; the other becomes arbitrarily long. Since

$$\lim_{n\to\infty} \rho \left(\right) = \tau^{1/2} + \tau^{-1/2},$$

this cannot occur either.

It follows that the set of limit points we seek are the numbers $\rho(S_{n+2}, v)$, where S_n is a path of length n, v a vertex adjacent to an end of S_{n+2}.

5. **Completion of proof.** As we have remarked before, we must find the largest root of $\theta P_{n+2} - Q$, where P_m is the characteristic

polynomial of a path with m vertices, $Q = xP_n$, $\theta = (x + \sqrt{x^2-4})/2$.
One readily calculates $P_n = \frac{1}{\theta - 1/\theta} (\theta^{n-1} - \theta^{-(n+1)})$. From this, using
$x = \theta + 1/\theta$, $z = \theta^2$, we seek the largest root of

(5.1) $$z^{n+1} - (1 + z + \cdots + z^{n-1}).$$

This proves the first part of the theorem. Next, we must show that
if β_n is the largest root of (5.1), $\beta_n \to \tau$. But this follows from
Proposition 3.6 or can easily be seen from (5.1) directly.

6. **Remarks.** It can be shown that $\tau^{1/2} + \tau^{-1/2}$ is also the
limit point of limit points of limit points of ... R. It can also
be shown that if G is a connected graph and not a tree, then $\rho(G)$
is a limit point. But I do not know if every limit point is an
algebraic number, for instance. Possibly for some λ, every number
at least λ is a limit point.

On least eigenvalues, I can find all limit points > -2 of least
eigenvalues of graphs (and these are algebraic integers), but I know
nothing about the range < -2. And I know nothing at all about limit
points for eigenvalues other than the greatest and least.

WHICH GENERALIZED PRISMS ADMIT H-CIRCUITS?

Victor Klee
University of Washington
Seattle, WA 98105
and
IBM Watson Research Center
Yorktown Heights, NY 10598

An H-path (H-circuit) for a graph G is a simple path (circuit) in G which involves all of G's nodes. If a graph admits an H-circuit then it admits a number of H-paths, but the converse is false. The generalized prisms, defined below, are of interest as a class of graphs in which H-paths always exist but the existence of H-circuits is problematical. They are related to the generalized Petersen graphs studied by Bondy [1], Castagna and Prins [2], Robertson [3], and Watkins [4].

By an n-permutation we mean a permutation $\pi = (\pi(1), \ldots, \pi(n))$ of $(1, \ldots, n)$. For each n-permutation π with $n \geq 3$, let $G(\pi)$ denote the trivalent graph consisting of a circuit (called the x-circuit) with successive nodes $x_1, \ldots, x_n$, another circuit (called the y-circuit) with successive nodes $y_1, \ldots, y_n$, and additional edges joining x_i to $y_{\pi(i)}$ for $1 \leq i \leq n$. That is, $G(\pi)$ consists of the $2n$ nodes $x_1, \ldots, x_n, y_1, \ldots, y_n$ and the $3n$ undirected edges $\{x_1, x_2\}, \{x_2, x_3\}, \ldots, \{x_n, x_1\}, \{y_1, y_2\}, \{y_2, y_3\}, \ldots, \{y_n, y_1\}, \{x_1, y_{\pi(1)}\}, \{x_2, y_{\pi(2)}\}, \ldots, \{x_n, y_{\pi(n)}\}$. For example, $G(1,2,3,4,5)$ is the ordinary pentagonal prism and $G(2,4,1,3,5)$ is the well-known Petersen graph.

A graph of the form $G(\pi)$ is here called a generalized prism or, more precisely, a generalized n-prism. It admits at least $4n$ H-paths, since for each edge $\{x_i, y_{\pi(i)}\}$ there are four distinct H-paths using that edge along with all but one edge of the x-circuit and all but one edge of the y-circuit. We are concerned here with the following two questions of Ralph Willoughby:

(Q1)   Which generalized prisms admit an H-circuit?

(Q2)   For which n does there exist a generalized n-prism not admitting an H-circuit?

Neither of the above questions is answered completely, but the following two results are established.

THEOREM 1. <u>For</u> n ≤ 8, <u>the only generalized n-prisms not admitting an H-circuit are those isomorphic with the Petersen graph</u>.

THEOREM 2. <u>For odd</u> n ≥ 3, <u>there exists a generalized n-prism not admitting an H-circuit if and only if</u> n <u>is neither</u> 3 <u>nor</u> 7.

For n ≡ 5 (mod 6), the existence of a generalized n-prism not admitting an H-circuit follows from results of Robertson [3] and Bondy [1] on generalized Petersen graphs.

I am indebted to Ralph Willoughby and Richard Guy for comments concerning the problem, and to Harlan Crowder and Philip Wolfe for programming suggestions.

Theorem 1 was established by a computer search over a reduced set of permutations. Let us say that two members j and k of {1, ..., n} are <u>n-adjacent</u> provided that |j-k| ∈ {1,n-1}, and that the n-permutation π is <u>obviously Hamiltonian</u> provided that there exist n-adjacent j and k in {1, ..., n} for which π(j) and π(k) are also n-adjacent. If π is obviously Hamiltonian, then G(π) admits an H-circuit formed from {x_j,y_{π(j)}}, {x_k,y_{π(k)}}, all edges of the x-circuit except {x_j,x_k}, and all edges of the y-circuit except {y_{π(j)},y_{π(k)}}. It suffices, therefore, to consider the n-permutations π that are not obviously Hamiltonian, and they may be <u>normalized</u> (by relabeling of the y_i's) so that π(n) = n and π(1) < π(2). When n is 3 or 4, all n-permutations are obviously Hamiltonian. The only normalized 5-permutation that is not obviously Hamiltonian is (2,4,1,3,5). It corresponds to the Petersen graph, which is easily seen not to admit an H-circuit. For n = 6, 7, and 8, the number of normalized n-permutations that are not obviously Hamiltonian are respectively 4, 29, and 216. By means of a computer program, these permutations were generated and the corresponding generalized prisms tested for H-circuits. In each case, an H-circuit was found. As a check, the test for H-circuits was also carried out in a different way, which is described later.

The notion of a "bad permutation", to be defined shortly, is basic in all that follows. For each m-permutation π with m ≥ 3, let G'(π) denote the subgraph of G(π) formed by omitting the edges {x_m,x_1} and {y_m,y_1}. For example, G'(1,2,3,4) and G'(2,4,1,3) are shown in Figure 1 below. The graph G'(π) is also defined in the obvious way when m is 1 or 2.

By an <u>E-path</u> in G'(π) we mean a simple path whose ends are both in the set E = {x_1,x_m,y_1,y_m}. An <u>EH-path</u> (<u>EH-pair</u>) is an E-path (pair of disjoint E-paths) involving all nodes of G'(π). An EH-path

175

is called <u>level-preserving</u> if its ends are both x's or both y's;
otherwise, it is <u>level-reversing</u>; the same terms are defined in the
natural way for EH-pairs. The permutation π is called <u>good</u> provided
that $G'(\pi)$ admits a level-reversing EH-path or EH-pair; otherwise,
π is <u>bad</u>. It is easily verified that $(2,4,1,3)$ is bad.

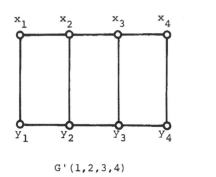

$$G'(1,2,3,4)$$

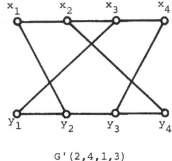

$$G'(2,4,1,3)$$

Figure 1.

PROPOSITION 1. <u>If</u> π <u>is an n-permutation with</u> $\pi(n) = n \geq 3$,
<u>and</u> $\bar{\pi}$ <u>is the restriction of</u> π <u>to</u> $\{1,\ldots,n-1\}$, <u>then</u> $G(\pi)$ <u>admits</u>
<u>an H-circuit if and only if the (n-1)-permutation</u> $\bar{\pi}$ <u>is good</u>.

<u>Proof.</u> Let $G'(\pi)$ be embedded in the natural way as a subgraph
of $G(\pi)$. If $\bar{\pi}$ is good then $G'(\bar{\pi})$ admits a level-reversing EH-
path or EH-pair, and this path or pair of paths is easily extended to
an H-circuit of $G(\pi)$. Conversely, if $G(\pi)$ admits an H-circuit C
then C intersects $G'(\bar{\pi})$ in a level-reversing EH-path or a level-
reversing EH-pair according as the edge $\{x_n, y_n\}$ appears or does not
appear in C.

COROLLARY. <u>There exists a generalized n-prism not admitting an</u>
<u>H-circuit if and only if there exists a bad (n-1)-permutation</u>.

With the aid of Proposition 1, Theorem 1 is seen to be equivalent
to the assertion that for $n \leq 7$, the only bad n-permutations are
$(2,4,1,3)$ and its reversal $(3,1,4,2)$. The latter assertion was
established directly by means of a computer program for generating all

n-permutations π such that (a) $\pi(1) < \pi(2)$, (b) $\{1,n\} \cap \{\pi(1), \pi(n)\} = \emptyset$, and (c) $1 \leq m < n \Rightarrow m < \max\{\pi(i): 1 \leq i \leq m\}$, and testing the associated graphs $G'(\pi)$ for EH-paths. The program was run for $4 \leq n \leq 7$, and each of the tested permutations other than $(2,4,1,3)$ turned out to admit more than two pairs in E as the end-pairs of EH-paths. The assertion about bad n-permutations for $n \leq 7$ then follows easily.

An n_1-permutation π_1 and an n_2-permutation π_2 can be catenated to form an (n_1+n_2)-permutation $\pi = (\pi_1, \pi_2)$ defined as follows: $\pi(i) = \pi_1(i)$ for $1 \leq i \leq n_1$; $\pi(i) = n_1 + \pi_2(i-n_1)$ for $n_1 < i \leq n_1 + n_2$. The operation of catenation is extended in the natural way to an arbitrary finite sequence of permutations.

PROPOSITION 2. Suppose that π_i is an n_i-permutation for $1 \leq i \leq k$, and that the following three conditions are all satisfied:

(a) for each i, either $n_i = 1$ or π_i is bad;

(b) there is an even (possibly zero) number of 1's among the n_i's;

(c) $1 < n_1$, $1 < n_k$, and no two 1's among the n_i's appear consecutively.

Then the catenate $\pi = (\pi_1, \ldots, \pi_k)$ is a bad permutation.

Proof. Let $n = \Sigma_1^k n_i$, and let the graphs $G'(\pi_i)$ be embedded in the natural way in the graph $G'(\pi)$. The subgraph $G'(\pi_i)$ will be called a large block or a small block according as π_i is bad (whence $n_i \geq 4$) or $n_i = 1$. Now consider an arbitrary EH-path P for $G'(\pi)$. If one end of P belongs to $\{x_1, y_1\}$ and the other to $\{x_n, y_n\}$, then P intersects each block in an EH-path for that block. Hence P is level-preserving across each large block and level-reversing across each small block, and since the total number of small blocks is even P is level-preserving relative to $G'(\pi)$. If, on the other hand, both ends of P are in $\{x_1, y_1\}$, then the intersection of P with $G'(\pi_n)$ is a level-reversing EH-path. A similar contradiction ensues if both ends of P are in $\{x_n, y_n\}$. Now suppose, finally, that $G'(\pi)$ admits a level-reversing EH-pair. If both of the paths in the pair stretch from $\{x_1, y_1\}$ to $\{x_n, y_n\}$, they intersect some large block in a level-reversing EH-pair for that block. If, on the other hand, one of the paths joins x_1 to y_1 and the other joins x_n to y_n, then, since no two small blocks appear consecutively, there is a large block that has connected intersection with both of the paths in the EH-pair. If both intersections are non-empty, they form a level-reversing EH-pair for the large block in question, while if one is empty the other forms a level-reversing EH-path for the large block. In each case, a contradiction ensues and the desired conclusion follows.

Except in the case $n = 11$, the part of Theorem 2 not covered by Theorem 1 is a consequence of Proposition 2, the Corollary of Proposition 1, and the existence of a bad 4-permutation. For $n = 11$, see the following paragraph.

Note that Willoughby's question (Q2) remains unanswered for all even $n \geq 10$, and that (Q1), which requests a direct characterization of those n-permutations π such that $G(\pi)$ admits an H-circuit, is unanswered for all $n \geq 9$. In connection with (Q1), there is a special interest in the n-permutations of the form $\pi_{m,n}$ given by

$$\pi_{m,n}(i) = \text{residue of } im \pmod n,$$

for $1 \leq i \leq n$, where m and n are relatively prime with $1 < m < n/2$ and where $\pi_{m,n}(n)$ is defined to be n. All such permutations were tested for $n \leq 12$, and the only ones not giving rise to H-circuits turned out to be $\pi_{2,5}$, $\pi_{2,11}$, and $\pi_{5,11}$. Actually, the last two give rise to isomorphic graphs [4], and their presence in this list completes the proof of Theorem 2.

I would not be suprised by the existence of a simple algorithm which, whenever n is odd and π is an n-permutation, would produce a level-reversing EH-path for the graph $G'(\pi)$. However, I am unable to prove that such paths always exist.

As I discovered only after completing the above investigation, some of the results included there can be proved in a different way, that is, by using the fact that $G(\pi_{m,n})$ is a generalized Petersen graph in the sense of Watkins [4]. A theorem of Robertson [3] implies that $G(\pi_{2,n})$ admits an H-circuit except when $n \equiv 5 \pmod 6$. The same was established by Bondy [1], who observed also that $G(\pi_{3,n})$ always admits an H-circuit $(n > 5)$. A conjecture of Castagna and Prins [2] implies that $G(\pi_{m,n})$ admits an H-circuit whenever $2 < m < (n-1)/2$. The existence of Tait cycles in generalized Petersen graphs was studied by Robertson [3], Watkins [4], and Castagna and Prins [2], and the property of being hypohamiltonian was studied by Bondy [1].

References

1. J. A. Bondy, Variations on the Hamiltonian theme, _Canad_. _Math_. _Bull_. 15 (1972), 57-62.

2. F. Castagna and G. Prins, Every generalized Petersen Graph has a Tait Coloring, _Pacific_ J. _Math_. 40 (1972), 53-58.

3. G. N. Robertson, _Graphs_ _under_ _Girth_, _Valency_, _and_ _Connectivity_ _Constraints_, Dissertation, University of Waterloo, Waterloo, Ontario, Canada, 1968.

4. M. E. Watkins, A Theorem on Tait Colorings with an Application to the Generalized Petersen Graphs, _J_. _Combinatorial_ _Theory_ 6 (1969), 152-164.

THE CHROMATIC NUMBER OF TRIANGLE-FREE GRAPHS

Hudson V. Kronk
SUNY at Binghamton
Binghamton, NY 13901

The chromatic number of triangle-free graphs has been the object of several investigations. Two of the most interesting results which have been obtained are stated below.

THEOREM A. For every positive integer k there exists a k-chromatic triangle-free graph.

THEOREM B. Every planar triangle-free graph has chromatic number not exceeding three.

The first proofs of Theorem A were published by Descartes [2] and Zykov [9]. Using probabilistic methods, Erdös [4] was able to show that for any two positive integers k and g, with k ≥ 2 and g ≥ 3, there exists a k-chromatic graph with girth greater than g. Theorem B is due to Grötzsch [5]. This note is concerned with the following question: What is the analogue of Grötzsch's Theorem for graphs having positive genus?

The case of toroidal graphs has been considered by Kronk and White [7] and is summarized in:

THEOREM C. Let G be a toroidal graph having girth g. Then

$$\chi(G) \leq 7, \text{ if } g = 3$$

$$\leq 4, \text{ if } g = 4 \text{ or } 5$$

$$\leq 3, \text{ if } g \geq 6.$$

Furthermore, all the bounds are sharp, except possibly for g = 5.

We now consider graphs having genus greater than one. We let $\Gamma = \frac{g}{g-2}$, where g denotes the girth.

THEOREM D. Let G be a k-chromatic graph having girth $g \geq 4$ and genus $\gamma \geq 2$. If $k \geq 1 + 2\Gamma$, then

$$k \leq \frac{1}{6}\left[3 + 6\Gamma + \sqrt{57 - 60\Gamma + 36\Gamma^2 + 48\Gamma\gamma}\right].$$

Proof. It is not difficult to show that if the bound holds when G is k-critical, then it holds in general. Hence we shall assume that G is k-critical; i.e., every proper subgraph of G has chromatic number less than k. It follows from the generalized Euler polyhedral formula (see, for example [6, p. 125]) that

(1) $$2q \leq 4\Gamma\gamma + 2\Gamma p - 4\Gamma,$$

where p and q denote the number of points and lines of G. Dirac [3] has shown that a k-critical, $k \geq 3$ triangle-free graph has

(2) $$2q \geq (k-1)p + 2k - 6.$$

From (1) and (2) we obtain

(3) $$(k-1-2\Gamma)p + 2k - 6 \leq 4\Gamma\gamma - 4\Gamma.$$

It is easily shown that for $k \geq 4$, a k-chromatic triangle-free graph has at least $3k - 2$ points. Since $k \geq 1 + 2\Gamma$, it follows from (3) that

(4) $$(k-1-2\Gamma)(3k-2) + 2k - 6 \leq 4\Gamma\gamma - 4\Gamma.$$

The desired bound is now obtained by solving the inequality (4) for k.

REMARKS. The bound given in Theorem D is probably far from best possible. It is not known whether the result of Dirac, which was used in the proof of Theorem D is best possible. Also, the bound $3k - 2$ on the number of points is not best possible. The author does not believe it is known how many points a k-chromatic triangle-free graph must have. Chvátal remarks in [1] that the smallest 4-chromatic triangle-free graph has 11 points and was constructed by Mycielski [8].

References

1. V. Chvátal, The smallest triangle-free 4-chromatic 4-regular graph, *J. Combinatorial Theory* 9 (1970), 93-94.

2. B. Descartes (pseudonym of Tutte), A three color problem, *Eureka* (April 1947); Solution (March 1948).

3. G. Dirac, A theorem of R. L. Brooks and a conjecture of H. Hadwiger, *Proc. London Math. Soc.* (3) 7 (1957), 161-195.

4. P. Erdös, Graph theory and probability, *Canad. J. Math.* 11 (1959), 34-38.

5. B. Grötzsch, Ein Driefarbensatz für dreikreisfreie Netze auf der Kugel, Wiss. Z. Martin - Luther Univ., Halle-Wittenberg, *Math. Nat. Reihe* 8 (1958.59), 109-120.

6. F. Harary, *Graph Theory*, Addison-Wesley, Reading, Mass., 1969.

7. H. Kronk and A. White, A 4-color theorem for toroidal graphs, *Proc. Amer. Math. Soc.*, to appear.

8. J. Mycielski, Sur le coloriage des graphs, *Colloq. Math.* 3 (1955), 161-162.

9. A. A. Zykov, On some properties of linear complexes, *Math. Sb.* 24 (1949), 163-188.

PLANARITY OF CAYLEY DIAGRAMS

Henry W. Levinson
Rutgers University
New Brunswick, NJ 08903

and

Elvira Strasser Rapaport
SUNY at Stony Brook
Stony Brook, NY 11790

1. Introduction. We formulate a method by which certain presentations of groups may be shown to have planar graphs. As we consider
arbitrary groups, we allow infinite graphs. We make use of certain
special embeddings we call point symmetric embeddings of Cayley diagrams (or Cayley graphs) in orientable surfaces. Subsequently certain
criteria are developed for planarity and the solvability of the word
problem. Finally we raise certain questions which may be accessible
to the techniques employed.

2. Conditions implying planarity. We begin this section with
some notation.

Let w be any cyclically reduced word in the generators of a
free group, F . By a short conjugate of w shall be meant any conjugate of w by a word $v \in F$ such that the length of $v^{-1}wv$, cyclically reduced, equals the length of w . A short conjugate of w
is therefore any word whose symbols are those of w cyclically
permuted.

We distinguish between embedded and unembedded graphs by using an
asterisk, L^* , to denote that a graph is unembedded, and its absence,
L , to denote that the graph in question is embedded.

Let $G \cong P = \langle X;R \rangle = F/N$ be a presentation on the symmetrized
sets X and R (X a finite set of symbols generating the free group
F , R a finite set of words in the symbols, both sets closed under
inversion and short conjugation). Let Γ be an Edmonds embedding [1]
of Γ^* , the Cayley diagram of P , in a compact orientable 2-manifold,
M , such that the counterclockwise succession of edges at each vertex
is the same. Call this sequence of edges S . Let $L^* \subseteq \Gamma^*$, the
local graph at v , (v an arbitrary fixed vertex) consist of v , all

edges incident with v, and all circuits corresponding to the $R_i \in R$
and incident with v. Let L be the embedding of L^* on M obtained from the restriction of Γ.

THEOREM 1. If S can be so chosen that L is planar, then Γ^*
is planar.

Proof. If C is a circuit of Γ and corresponds to the word w
in F (i.e. the edges of C are labeled with the $x_i^{\pm 1}$ so as to
spell out w), write $C \sim w$.

Let v be assigned to the identity element of F and write
$v \sim 1$. Then every vertex is assigned an element of G. Thus if v,v'
is a path ρ in Γ^*, starting at v and ending at v', corresponding to w, then $v' \sim w$. If w is a product of conjugates of
the $R_i \in R$, that is,

$$w = \prod_{j=1}^{k} z_j^{-1} R_{i_j} z_j,$$

then w is a circuit at $v \sim 1$, and

$$uwu^{-1} = \prod_{j=1}^{k} u z_j^{-1} R_{i_j} z_j u^{-1}$$

is a circuit at $v \sim 1$. The latter consists of a path, u (from v
to $v_o \sim u$), the circuit for w (from v_o to v_o), and the path
for u^{-1} (from v_o to v). Write w_1 for the path corresponding to
w and starting at $v \sim 1$ (in general, write w_u for the path
starting at u.

Since the R_i incident with v bound on M (i.e. are homol-
ogous 0 on M), and since S is independent of choice of v, we
have, as a consequence of the (prior use of the) Edmonds embedding
technique [1] that the R_i bound at each vertex. Thus if $w \equiv 1$ in
G and $C \sim w$, then for every $u \in F$, w_u is a circuit bounding on M.

Assume now that M is so chosen that its genus is least possible.
It may be the case that for some sequence S' of edges at a vertex
for which L is again planar, another surface of lower genus results.
Assume S was picked so this genus is minimal. Suppose then this
genus is not zero. Then there exists a circuit C in Γ that does

not bound on M. Let v_0 be a vertex on C. Then $C \sim w_u$, for some $w \in N$, $v_0 \sim u$, and w bounds on M, which is a contradiction.

We call an embedding of a Cayley diagram on a compact orientable 2-manifold point symmetric if the counterclockwise succession of edge labels is the same at every vertex, to within cyclic permutation.

For a presentation, P, let L_v^* be the local graph at v as defined above. Given some vertex, $v' \in L_v^*$, there may not be a local graph at v' for the presentation P, in L_v^*. If not, construct a copy, $L_{v,v'}^*$, of L_v^* and identify the vertex in $L_{v,v'}^*$, corresponding to v in L_v^*, with v' in L_v^*. Now there are pairs of identically labeled edges incident at v. Identify any two pairs of identically labeled edges incident at the same vertex until no two edges are incident at the same vertex unless their labels are different. We call this construction continuing L_v^* at v', and we note that this process may be iterated any number of times at vertices at which there is not yet a local graph for P. We further point out that should the process be attempted at a vertex at which there already is a local graph for P, no net change will result since the "added" local graph will merely be superimposed on the extant local graph by the identifications of identically labeled edges.

Theorem 2 gives the genus of the graph Γ^* of $P = \langle X; R \rangle$, in certain cases, without knowledge of L^* or Γ^*. Let P be symmetrized as before, and X and R contain n and m elements respectively. Consider the graph constructed in 3-space via the following four steps.

1. Start with a vertex v.

2. For each R_i in R, construct a distinct directed circuit C_i incident to v whose edges are labeled from v to correspond with R_i. (The only vertex two distinct C_i, C_j have in common at this step is v.)

3. If a path v_1, v, v_2, consisting of two edges, (v_1, v) and (v, v_2), corresponds to the word aa^{-1} or $a^{-1}a$, then v_1 is identified with v_2, and edge (v, v_1) with edge (v, v_2). This is repeated in the resulting graph at each vertex, one at a time, in any order, until no path aa^{-1} or $a^{-1}a$ remains.

Suppose there is an embedding of this graph in the plane E^2.

4. To each vertex (of the embedded graph) lacking x_i or x_i^{-1}, $i = 1, \ldots, n$, a properly directed edge labeled x_i is attached.

Call the result L_o, and its unembedded counterpart L_o^*.

THEOREM 2. If L_o can be chosen point symmetric, then Γ^* is planar.

Proof. We proceed by first continuing L_o^* and showing we get Γ^*, the Cayley diagram for P, and then embedding Γ^* in E^2. Continue L_o^* at each of its vertices. Clearly the result is unique, depending only on L_o^*. Every vertex of L_o^* now has n distinct edges incident to it. At each vertex there is a circuit corresponding to R_i in R, for each i = 1, ..., m. If now every circuit label (word) corresponds to a consequence of R and conversely, then the graph is Γ^*. We prove this for the graph so far constructed, but note that the proof is identical for the graph resulting from unlimited continuation.

Let e be an edge in L_o^* but not on a circuit. When L_o^* is continued, e will be on a circuit if and only if it is covered in the act of continuation by an edge e' which is on a circuit of some copy of L_o^* used in the act of continuation. Thus a circuit obtained by continuation contains no edge attached in step 4 (nor any copy of such used during the continuation). Let C be a circuit in the graph obtained by the continuation process such that C is neither in L_o^* nor in a copy, L_1^*, of L_o^* alone. Then $C = e_1 \ldots ee' \ldots e_k$; $e \in L_o^*$, $e' \in L_1^*$, and the vertex, v, between them is in both L_o^* and L_1^*. As $e \notin L_1^*$, v is not on a circuit in L_1^* and so v was attached in step 4, which is impossible. Hence in the continuation, all circuits are products of those already in L_o^*, or in copies of L_o^*.

For the converse, let the path ρ from v to v' have label corresponding to

$$R_i = \prod_{j=1}^{k} x_{i_j}^{e_j} .$$

Then C is labeled by $x_{i_1}^{e_1} x_{i_2}^{e_2} \ldots x_{i_k}^{e_k}$, where the $e_j = \pm 1$. As each edge label (from X) appears exactly once at each vertex, paths from any fixed vertex, v, are uniquely described by words in the elements of X. At v there is a circuit labeled $x_{i_1}^{e_1} x_{i_2}^{e_2} \ldots x_{i_k}^{e_k}$ (corresponding to R_i). Therefore we have v = v'.

Thus the graph obtained after unlimited continuation is Γ^*.

Moreover L_o^* is L^* of Theorem 1.

Hence L_o is L and if L is point symmetric, then Theorem 1 applies. Therefore Γ^* is planar.

Note that if in L_o all the circuits for the R_i bound finite faces (i.e. the edges attached in step 4 all end in the infinite face of the graph on hand), then Γ^* has an embedding in E^2 which is (point symmetric and) locally finite in the sense that there is no accumulation point of vertices in any finite part of E^2. For then continuation at every step takes place in the infinite face of the previously embedded graph.

Theorem 2 picks out the case (presentation) when the putative local graph L_o^* turns out to be a subgraph (the local graph, L^*) of Γ^*. Theorem 3 states the least amount of information needed to know the local graph L_o^*. (Theorem 3 dispenses with Γ^* in Theorem 1, i.e. delimits the need to know if for the word problem.)

Let P be a symmetrized finite presentation of the group G, as before, and let $R_i = x_{i_1}^{e_1} x_{i_2}^{e_2} \cdots x_{i_k}^{e_k}$, k depending on i. Let $\underline{a}$ be one of the x_i. Let

$$R_{i_1} = \underline{aa}^{-1} x_{i_1}^{e_1} \cdots x_{i_k}^{e_k}$$

$$R_{i_2} = x_{i_1}^{e_1} \underline{aa}^{-1} x_{i_2}^{e_2} \cdots x_{i_k}^{e_k}$$

$$\vdots$$

$$R_{i_k} = x_{i_1}^{e_1} \cdots x_{i_{k-1}}^{e_{k-1}} \underline{aa}^{-1} x_{i_k}^{e_k} \ .$$

Let $\overline{R}$ be the set of all such words as $\underline{a}$ ranges over X and R_i ranges over R. Let T be the set of all initial segments of elements of $\overline{R}$, with each member of T taken freely reduced.

THEOREM 3. Suppose it is known which pairs of elements of T are equal in G. Then L_o^* of Theorem 2 is known. (Hence Γ^* need not be known in order to use Theorem 1.)

Proof. Note that the premise is equivalent to knowing which vertices of the circuits for R, and of edges incident with them, coincide. So we have L_o^*.

The set $\overline{R}$ is needed to decide if two vertices on the graph constructed in steps 1 - 4 prior to Theorem 2 coincide in Γ^*.

COROLLARY. A presentation satisfying the premises of Theorems 2 or 3 has solvable word problem.

Proof. Solving the word problem and constructing Γ^* are equivalent.

The assumption in Theorem 2 which allows Γ^* to be constructed (which solves the word problem for P) is that L_o^* be point symmetric planar. Point symmetric planarity of L_o^* is decidable for finite presentations; however if L_o^* is not point symmetric planar, the test fails.

3. Conclusion. Use has been made of the notion of point symmetric embeddings and the bounding properties of Jordan curves in the plane. We may relax the notion as follows. An embedding of a Cayley diagram in an orientable compact 2-manifold is called weakly point symmetric if the succession of edges either clockwise or counterclockwise is the same at each vertex, to within a cyclic permutation (or an inverse of a cyclic permutation). Under the conditions of Theorem 1, with point symmetry replaced by weak point symmetry, the local graph L^* may be continued. Can it be embedded in the projective plane, or in some suitable "minimal" non-orientable compact 2-manifold? Since bounding properties of Jordan curves no longer apply in non-orientable surfaces, the problem of the projective planarity of a presentation with weakly point symmetric local graph remains.

References

1. J. Edmonds, A combinatorial representation for polyhedral surfaces, Notices Amer. Math. Soc. 7 (1960), 646.

HAMILTONIAN AND EULERIAN PROPERTIES OF
ENTIRE GRAPHS

John Mitchem
California State University
San Jose, CA 95114

In a plane graph G let $e_1 = uv$ be an edge. We say e_1 is
adjacent to vertices u and v, which are also adjacent to each
other. Also e_1 is adjacent to edge $e_2 = uw$. A face of G is adjacent to the vertices and edges which are on its boundary, and two
faces of G are adjacent if their boundaries share a common edge.
The entire graph of G, denoted e(G), is the graph with vertex set
the vertices, edges, and faces of G. Two vertices of e(G) are
adjacent if and only if they are adjacent elements of G.

In [2] H. Izbicki reported that M. Neuberger, by assuming the
four color conjecture, has shown that for any bridgeless cubic plane
graph G, the chromatic number of e(G) is at most seven. Kronk and
Mitchem, in [3], have proved the same result without using the four
color conjecture. In this paper we develop a necessary and sufficient
condition for the entire graph of a connected plane graph to be
eulerian, as well as investigate hamiltonian properties of entire
graphs. Also we show exactly which graphs have planar entire graphs.

We first observe that plane graphs G and H may be isomorphic,
but e(G) and e(H) may have different properties, because G and
H are embedded differently in the plane. For example in Figure 1,
the chromatic number of e(G) is 6, and the chromatic number of
e(H) is 5.

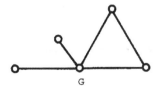
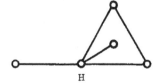

G H

Figure 1.

Before proceeding with necessary and sufficient conditions for a connected plane graph G to have an eulerian entire graph, we make the following remarks.

REMARK 1.   For v a vertex of a plane graph G, $\deg_{e(G)} v = 2 \deg_G v + r$, where r is the number of faces adjacent to v.

REMARK 2.   Let x = uv be an edge of a plane graph G; then $\deg_{e(G)} x = \deg_G u + \deg_G v + r$, where r = 1, if x is a bridge, and r = 2 otherwise.

REMARK 3.   Let R be a cyclic face of a plane graph G. Then $\deg_{e(G)} R = 2n + r$, where n is the number of vertices in the cycle which borders R and r is the number of faces adjacent to R.

THEOREM 1.   If G is a connected plane graph with a bridge, then e(G) is not eulerian.

Proof. Suppose e(G) is eulerian. Edge $x = w_1 w_2$ of G has even degree in e(G). Remark 2 implies that $\deg_G w_1$ and $\deg_G w_2$ have opposite parity if and only if x is a bridge. Let x_1 be a bridge of G such that one of the components, call it C, of $G - x_1$ contains no bridges. Let $x_1 = uv$, where $v \in V(C)$. Since $\deg_{e(G)} v$ is even, Remark 1 implies that C must contain a cycle. We consider two cases depending on the parity of v in G.

Case i. $\deg_G v$ is odd. The graph C must have two vertices which are not cutvertices of C. Let $v_1 \neq v$ be one of these two vertices, and thus v_1 is not a cutvertex of G. Let $v_1, v_2, \ldots,$ $v_s = v$, $s \geq 2$, be a path in C. For i = 1, ..., s - 1, edge $v_i v_{i+1}$ is not a bridge of G and thus, by Remark 2, $\deg_G v_i$ and $\deg_G v_{i+1}$ have the same parity. This implies that $\deg_G v_1$ is odd, say 2b + 1. However, since v_1 is not a cutvertex, it is adjacent to exactly 2b + 1 faces. According to Remark 1 $\deg_{e(G)} v_1$ is odd, which is a contradiction.

Case ii. $\deg_G v$ is even. Since $x_1 = uv$ is a bridge, the vertex $v = v_1$ is contained in a block B_1 of C such that $\deg_{B_1} v_1$ is odd. There is a vertex $v_2 \neq v_1$ such that $\deg_{B_1} v_2$ is odd. Since no edge of C is a bridge of G, the degree in G of each vertex of C is even. All adjacencies of v_2 are in C which implies that v_2 is in block B_2 (different from B_1) and $\deg_{B_2} v_2$ is odd. There is a vertex v_3 such that $\deg_{B_2} v_3$ is odd. Thus v_3 is also in block B_3, different from B_1 and B_2 and $\deg_{B_3} v_3$ is odd. This procedure must continue without end. However, this is impossible since the graph is finite.

THEOREM 2. If G is a connected plane graph and e(G) is eulerian, then G is eulerian.

Proof. Suppose a connected plane graph G is noneulerian, but e(G) is eulerian. Then G has a vertex of odd degree and, according to Theorem 1, G is bridgeless. Remark 2 implies that all vertices of G have odd degree. For $v \in V(G)$, $\deg_{e(G)} v$ is even, and thus the number of faces of G adjacent to v is even. This implies that v is a cutvertex. That is, each vertex of G is a cutvertex, which is impossible.

THEOREM 3. Let G be a connected plane graph. A necessary and sufficient condition for e(G) to be eulerian is that each of the following hold:

i. G is eulerian.

ii. The number of faces adjacent to each vertex of G is even.

iii. Each face of G has an even number of elements adjacent to it.

Proof. The necessity is immediate; we consider the sufficiency. Let x be a vertex of e(G). If x is a vertex or face of G, then properties ii and iii imply that $\deg_{e(G)} x$ is even. Suppose x is an edge of G. Since G is eulerian, x = uv is not a bridge. This together with the fact that $\deg_G u$ and $\deg_G v$ are even imply that $\deg_{e(G)} x$ is even and e(G) is eulerian.

If graph G in Theorem 3 has no cutvertices, then each face R of G has a boundary which is a cycle. Then Remark 3 implies that $\deg_{e(G)} R = 2n + r$, where r is the number of faces adjacent to R. Thus condition iii can be replaced by:

iii'. Each face of G has an even number of faces adjacent to it.

We note that graphs G_1 and H_1 in Figure 2 are isomorphic. However, $e(G_1)$ is eulerian and $e(H_1)$ is not.

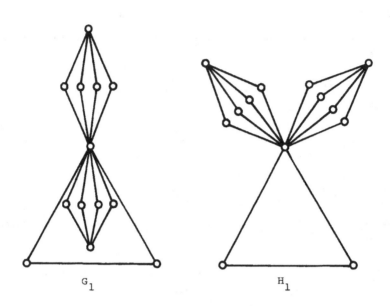

G_1 H_1

Figure 2.

We begin our consideration of hamiltonian entire graphs with the following theorem.

THEOREM 4. <u>Let</u> G <u>be a plane graph and</u> $e(G)$ <u>be hamiltonian.</u> <u>Then</u> G <u>is connected and no face of</u> G <u>has a boundary which contains five bridges with a common end vertex.</u>

<u>Proof.</u> Suppose G is not connected and let R be a face of G whose boundary contains vertices from two components of G. In $e(G)$, R is a cutvertex, which implies that $e(G)$ is not hamiltonian.

Assume G has a face, call it R, whose boundary contains bridges uv_1, uv_2, uv_3, uv_4, and uv_5. Let the maximal connected subgraph of G containing v_i but not u be denoted by G_i, $i = 1$, ..., 5. Suppose that $e(G)$ has a hamiltonian cycle C. The only elements of $G-G_i$ adjacent to elements of G_i are u, uv_i, and R. Thus in C the predecessor of the first element of G_i and the successor of the last element of G_i must be two of the three elements u, uv_i, and R. Since in C, u and R may each be adjacent to elements from at most two G_i, one of the G_i, say G_1, has no elements adjacent in C to u or R. However, this implies that the elements of G_1 do not lie on C, a contradiction.

It follows from Theorem 4 that the plane graph G_2 in Figure 3 does not have a hamiltonian entire graph. However, it is easily verified that $e(H_2)$ is hamiltonian.

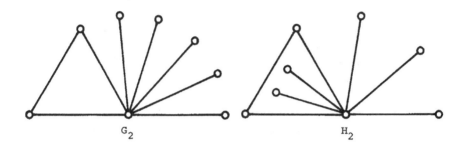

G_2 H_2

Figure 3.

We now give two classes of plane graphs whose entire graphs are hamiltonian.

THEOREM 5. If G is a hamiltonian plane graph, then e(G) is also hamiltonian.

Proof. Let v_1, e_1, v_2, ..., e_{n-1}, v_n, e_n, v_1 be a hamiltonian cycle of G. We form a hamiltonian cycle of $e(G)$. Start with v_1; then use the face r_1, adjacent to e_1 but clockwise from e_1. By selecting the edges and faces consecutively in a clockwise manner about v_1 from r_1 eventually we select e_1 and then proceed to v_2. Starting with the region clockwise adjacent to e_2, select consecutively, in a clockwise manner, the regions and edges adjacent to v_2 which have not already been used. Eventually e_2 is encountered followed by v_3. Continue this procedure until we encounter v_n and follow that with v_1. The resulting sequence v_1, r_1, ..., e_n, ..., e_1, v_2, ..., e_{n-1}, v_n, v_1 is a hamiltonian cycle of $e(G)$.

THEOREM 6. <u>Let</u> G <u>be a connected</u> <u>bridgeless</u> <u>plane</u> <u>graph</u> <u>such</u> <u>that each face of</u> G <u>is adjacent to a vertex of degree</u> 2 <u>or</u> 3. <u>Then</u> e(G) <u>is hamiltonian.</u>

<u>Proof.</u> In [1] Fleischner proved that the total graph of G is hamiltonian. Since e(G) contains the total graph of G, there is a cycle of e(G) which contains all vertices and edges of G. Let C be the largest such cycle in e(G). We show that C must contain each face of G.

Suppose C does not contain face R of G. If face R is adjacent to a vertex v of degree two, then R can be inserted in C between v and its successor in C, thus lengthening C, a contradiction. So the vertex u, of smallest degree adjacent to R, must have degree three. Let $e_1, e_2, \ldots, e_n$ be edges adjacent to R which form a cycle and such that e_1 and e_2 are both adjacent to u. If any two elements of G adjacent to R are consecutive in C, then R can be inserted between them forming a larger cycle than C. This implies that each edge e_i must be adjacent in C to edges x_i, y_i which are not adjacent to R. Let the edge adjacent to both e_1 and e_2 be e and consider three cases.

<u>Case</u> i. Edge e is adjacent in C to exactly one of e_1 and e_2, say e_2; then e_1, x_1, y_1 must have common vertex u_1. In C remove e_1 from between x_1 and y_1 and insert e_1 and R as $\ldots, e_2, R, e_1, e$. This forms a larger cycle than C.

<u>Case</u> ii. Edge e is adjacent in C to both e_1 and e_2. Then u must be adjacent in C to an element x which is also adjacent to R. Thus by inserting R between u and x we lengthen cycle C.

<u>Case</u> iii. Edge e is adjacent in C to neither e_1 nor e_2. Then there exist two adjacent edges, say e_i and e_{i+1}, in the cycle about R with the property that $x_i, y_i, x_{i+1}, e_i, e_{i+1}$ have a common vertex w. In C remove e_i and then insert e_i and R as $\ldots, y_{i+1}, e_{i+1}, R, e_i, x_{i+1}$. This again enlarges cycle C and completes the proof of the theorem.

We make the following conjecture which generalizes both Theorems 5 and 6. If C has no point of degree three or more, then C is a cycle and e(G) contains K_5.

CONJECTURE. If G is a bridgeless plane graph, then e(G) is hamiltonian.

We close with the following observation characterizing graphs which have planar entire graphs.

THEOREM 7. For a plane graph G a necessary and sufficient condition for e(G) to be planar is that G is a path or a union of disconnected paths.

Proof. Clearly if G is a path or the union of disconnected paths, then e(G) is planar. Suppose then that G has a component C which is not a path. Let v be a vertex of C of degree $n \geq 3$. If $n \geq 4$, then v and n edges are mutually adjacent in e(G). This implies that e(G) is non-planar since it contains the graph K_5. If n = 3 and v is a cutvertex, then there is a face R mutually adjacent with v and its three adjacent edges, so again e(G) contains K_5. If n = 3 and v is not a cutvertex, then e(G) contains K(3,3).

References

1. H. Fleischner, On Spanning Subgraphs of a Connected Bridgeless Graph and Their Application to DT-Graphs, to appear.

2. H. Izbicki, Verallgemeinerte Farbenzahlen, Beitrage zur Graphentheorie, (H. Sachs, H. Voss, and H. Walther, eds.), Teubner, Leipzig, 1968, 81-84.

3. H. Kronk and J. Mitchem, The Entire Chromatic Number of a Normal Graph is at Most Seven, Bull. Amer. Math. Soc., to appear.

TWO PROBLEMS ON RANDOM TREES

J. W. Moon
The University of Alberta
Edmonton, Alberta, Canada

1. **Introduction.** Let T denote a tree with n labeled nodes.
(For definitions not given here see [2] or [6].) The cutting number
of a node p in T is defined by Harary and Ostrand [3] to be the
number c_p of unordered pairs of nodes u and v such that $p \neq u,v$
and the path joining u and v in T passes through p.

If $1 \leq i < n$, let T_i denote the subtree of T determined by
those nodes j such that the path joining nodes n and j in T
passes through node i. The **mass** of T_i is the number m_i of nodes
in T_i. (The parameters m_i appear in some formulas in [7] that per-
tain to random walks on random trees.)

In sections 2 and 3 we shall determine the first two moments of
m_i and c_p over all the n^{n-2} trees T; in section 4 we shall show
how these two problems are both related to a third problem and hence
are related to each other in a sense.

The asymptotic behaviour of the formulas we shall obtain can be
determined by using the inequalities $x < -\ln(1-x) < x/(1-x)$, where
$0 < x < 1$, and then approximating the resulting sums by an appropriate integral of the type $\int_0^\infty x^k e^{-x^2} dx$. We shall omit the details of
the proofs of these asymptotic results as they are fairly routine.

2. **The expected mass of subtrees of a tree.** In the discussion
that follows we shall use the identities

$$\frac{1}{n^{n-2}} \sum_{v=0}^{n-2} \binom{n-2}{v} (v+1)^v (n-1-v)^{n-2-v} = \frac{n}{n-1} \sum_{t=2}^{n} \frac{(n)_t}{n^t} , \tag{1}$$

$$\frac{1}{n^{n-2}} \sum_{v=0}^{n-2} \binom{n-2}{v} (v+1)^{v+1} (n-1-v)^{n-2-v} = \frac{n}{n-1} \sum_{t=2}^{n} \binom{t}{2} \frac{(n)_t}{n^t} , \tag{2}$$

and

$$(3) \qquad \sum_{t=2}^{n} \binom{t}{2} \frac{(n)_t}{n^t} = \frac{1}{2} n \sum_{t=2}^{n} \frac{(n)_t}{n^t} .$$

Equations (1) and (2) are special cases of some Abel identities proved by Riordan [8; p. 23], and equation (3) follows from identities proved in [5]. We now determine the expected value of m_i and m_i^2 over all trees T with n labeled nodes.

THEOREM 1. If $n \geq 2$, then

$$(4) \qquad E(m_i) = \frac{n}{n-1} \sum_{t=2}^{n} \frac{(n)_t}{n^t}$$

and

$$(5) \qquad E(m_i^2) = \frac{1}{2} n \ E(m_i) .$$

COROLLARY 1. The expected values $E(m_i)$ and $E(m_i^2)$ satisfy the relations $E(m_i) \sim (\frac{1}{2}\pi n)^{1/2}$ and $E(m_i^2) \sim (\frac{1}{8}\pi)^{1/2} n^{3/2}$ as $n \to \infty$.

Proof. We first determine the number of trees T such that $m_i = k$, where $1 \leq k \leq n-1$. There are $\binom{n-2}{k-1}$ ways to choose the $k-1$ nodes of T_i other than i and, having chosen these nodes, there are k^{k-2} ways to form the tree T_i; there are $(n-k)^{n-k-2}$ ways to form a tree on the $n-k$ nodes not in T_i and if we now join i to any one of these $n-k$ nodes we obtain a tree T with the required property. Hence, there are

$$\binom{n-2}{k-1} k^{k-2} (n-k)^{n-k-1}$$

trees T such that $m_i = k$. Consequently,

$$E(m_i) = \frac{1}{n^{n-2}} \sum_{k=1}^{n-1} \binom{n-2}{k-1} k^{k-1} (n-k)^{n-k-1}$$

$$= \frac{1}{n^{n-2}} \sum_{v=0}^{n-2} \binom{n-2}{v} (v+1)^v (n-1-v)^{n-2-v} \ .$$

Formula (4) now follows from identity (1).

Similarly, we find that

$$E(m_i^2) = \frac{1}{n^{n-2}} \sum_{k=1}^{n-1} \binom{n-2}{k-1} k^k (n-k)^{n-k-1}$$

$$= \frac{1}{n^{n-2}} \sum_{v=0}^{n-2} \binom{n-2}{v} (v+1)^{v+1} (n-1-v)^{n-2-v} \ .$$

Formula (5) now follows from identities (2) and (3) and from formula (4).

3. **The expected cutting number of a node.** We may restrict our attention to the cutting number of node n. The definition of c_n implies that

$$c_n = \frac{1}{2} \sum' m_i m_j \ , \tag{6}$$

where in the summation, here and elsewhere, the subscripts range independently over all nodes that are adjacent to n except that different subscripts are not to assume the same value simultaneously. (We remark that the formula

$$c_n = \frac{1}{2} \sum' m_i (n-1-m_i) = \frac{1}{2}(n-1)^2 - \frac{1}{2} \sum' m_i^2$$

is perhaps more convenient for computational purposes.)

In the following discussion we shall use the facts that if

$$Y = \sum_{n=1}^{\infty} n^{n-1} \frac{x^n}{n!} \ ,$$

then

(7)
$$Y = xe^Y$$

and

(8)
$$\frac{y^t}{t!} = \sum_{n=t}^{\infty} \binom{n}{t} tn^{n-t-1} \frac{x^n}{n!}$$

for $t = 1, 2, \dots$. These relations can be proved in various ways; see, for example, [6; pp. 15 and 26]. We now desire a formula for $\mu(n)$, the expected value of c_n over all trees T with n labeled nodes.

THEOREM 2. If $n \geq 3$, then

(9)
$$\mu(n) = \frac{1}{2} n \sum_{t=3}^{n} \frac{(n)_t}{n^t} .$$

COROLLARY 2. The expected value $\mu(n)$ of c_n satisfies the relation $\mu(n) \sim (\frac{1}{8}\pi)^{1/2} n^{3/2}$ as $n \to \infty$.

Proof. If the n-th node of T is joined to k other nodes, then T can be formed from a forest of k rooted trees involving a total of $n - 1$ nodes by joining the roots of these trees to the n-th node. If

$$M = \sum_{n=3}^{\infty} \mu(n) n^{n-1} \frac{x^n}{n!} ,$$

then it follows from the preceeding observation and equation (6) that

$$M = x \sum_{k=2}^{\infty} \binom{k}{2} \frac{y^{k-2} (xY')^2}{k!} .$$

The trees T are classified according to the number k of edges incident with the root node; the product $Y^{k-2}(xY')^2$ is the generating function for the number of forests of k trees where the number of times each such forest is counted equals the product of the number of nodes in two particular subtrees in the forest; there are $\binom{k}{2}$ ways to specify the two particular subtrees; the $k!$ takes into account the symmetry between the trees in the forest, and the factor x takes into account the root node of T that is joined to the roots of the subtrees in the forest.

Equation (7) implies that $xY' = Y(1-Y)^{-1}$; hence,

$$M = \frac{1}{2} x e^Y Y^2 (1-Y)^{-2} = \frac{1}{2} Y^3 (1-Y)^{-2} = \frac{1}{2} \sum_{t=3}^{\infty} (t-2) Y^t.$$

If we equate coefficients of x^n in this relation, using (8), we obtain the formula

$$\mu(n) = \frac{1}{2} \sum_{t=3}^{n} t(t-2) \frac{(n)_t}{n^t}.$$

Formula (9) now follows from identity (3) and the fact that

$$\sum_{t=1}^{n} t \frac{(n)_t}{n^t} = n.$$

Clarke [1] proved that there are $\binom{n-2}{k-1}(n-1)^{n-k-1}$ trees T with n labeled nodes in which node n is joined to exactly k other nodes; let $\mu(n,k)$ denote the expected value of c_n over the set of such trees.

THEOREM 3. If $2 \le k \le n-1$, then

$$\mu(n,k) = \frac{1}{2}(k-1)(n-1) \sum_{t=0}^{n-1-k} \frac{(n-1-k)_t}{(n-1)^t}. \tag{10}$$

COROLLARY 3. If $k = o(n^{1/2})$, then $\mu(n,k) \sim (\frac{1}{8}\pi)^{1/2}(k-1)n^{3/2}$ as $n \to \infty$.

Proof. If

$$M_k = \sum_{n=k+1}^{\infty} \mu(n,k)\binom{n-2}{k-1}\left((n-1)^{n-k-1}\right)\frac{x^n}{(n-1)!} \ ,$$

then it follows by essentially the same argument as before that

$$M_k = \frac{x}{k!}\binom{k}{2}Y^{k-2}(xY')^2 = \frac{1}{2}\frac{x}{(k-2)!}Y^k(1-Y)^{-2} = \frac{1}{2}\frac{x}{(k-2)!}\sum_{t=0}^{\infty}(t+1)Y^{k+t} \ .$$

If we equate coefficients of x^n in this relation, we obtain the formula

$$\mu(n,k) = \frac{1}{2}(k-1)\sum_{t=0}^{n-k-1}(t+1)(t+k)\frac{(n-1-k)_t}{(n-1)^t} \ .$$

Formula (10) now follows from this and two identities given in [7].

Let $\gamma(n)$ denote the expected value of c_n^2 over all trees T with n labeled nodes.

THEOREM 4. If $n \geq 3$, then

(11) $$\gamma(n) = \sum_{t=3}^{n}\left\{\frac{1}{2}\binom{t+2}{5} + \binom{t}{4} + \frac{1}{4}\binom{t-2}{3}\right\}t\,\frac{(n)_t}{n^t} \ .$$

COROLLARY 4. The expected value $\gamma(n)$ of c_n^2 satisfies the relation $\gamma(n) \sim \frac{1}{32}(2\pi)^{1/2}n^{3\,1/2}$ as $n \to \infty$.

Proof. If we square both sides of equation (6), rearrange slightly, and take expectations, we find that

$$\gamma(n) = \frac{1}{2}\sum{}' E\{m_i(m_i-1)m_j(m_j-1)\} + \sum{}' E\{m_i(m_{i-1})m_j\} + \frac{1}{2}\sum{}' E\{m_im_j\}$$

$$+ \sum{}' E\{m_i(m_i-1)m_jm_h\} + \sum{}' E\{m_im_jm_h\} + \frac{1}{4}\sum{}' E\{m_im_jm_hm_\ell\} \ .$$

If

$$G = \sum_{n=3}^{\infty} Y(n)n^{n-1}\frac{x^n}{n!} ,$$

then it follows, by the same type of argument as was used in the proof of Theorem 2, that

$$G = \tfrac{1}{2}Y(x^2Y'')^2 + Y(x^2Y'')(xY') + \tfrac{1}{2}Y(xY')^2$$

$$+ Y(x^2Y'')(xY')^2 + Y(xY')^3 + \tfrac{1}{4}Y(xY')^4 .$$

If we now replace xY' by $Y(1-Y)^{-1}$, x^2Y'' by $Y^2(1-Y)^{-2} + Y^2(1-Y)^{-3}$, and then expand the resulting expression in powers of Y, we eventually find that

$$G = \sum_{t=3}^{\infty} \left\{\tfrac{1}{2}\binom{t+2}{5} + \binom{t}{4} + \tfrac{1}{4}\binom{t-2}{3}\right\}Y^t .$$

If we equate powers of x^n in this relation we obtain formula (11).

Let $Y(n,k)$ denote the expected value of c_n^2 over all trees T with n labeled nodes in which nodes n is joined to exactly k other nodes.

THEOREM 5. If $2 \le k \le n-1$, then

$$Y(n,k) = \tfrac{1}{2}(k-1) \sum_{t=0}^{n-k-1} \left\{\binom{t+5}{5} + 2(k-2)\binom{t+4}{4}\right. \tag{12}$$

$$\left. + \binom{k-2}{2}\binom{t+3}{3}\right\}(k+t)\frac{(n-1-k)_t}{(n-1)^t} .$$

COROLLARY 5. If $k = o(n^{1/2})$, then $Y(n,k) \sim \tfrac{1}{32}(2\pi)^{1/2}(k-1)n^{7/2}$ as $n \to \infty$.

<u>Proof.</u> If

$$G_k = \sum_{n=k+1}^{\infty} \gamma(n,k)\binom{n-2}{k-1}(n-1)^{n-k-1}\frac{x^n}{(n-1)!} \, ,$$

then it follows, by the same type of argument as before, that

$$G_k = \frac{1}{2}\frac{x}{(k-2)!}Y^{k-2}(x^2Y'')^2 + \frac{x}{(k-2)!}Y^{k-2}(x^2Y'')(xY')$$

$$+ \frac{1}{2}\frac{x}{(k-2)!}Y^{k-2}(xY')^2 + \frac{x}{(k-3)!}Y^{k-3}(x^2Y'')(xY')^2$$

$$+ \frac{x}{(k-3)!}Y^{k-3}(xY')^3 + \frac{1}{4}\frac{x}{(k-4)!}Y^{k-4}(xY')^4$$

$$= \frac{1}{2}\frac{x}{(k-2)!}\sum_{t=0}^{\infty}\left\{\binom{t+5}{5} + 2(k-2)\binom{t+4}{4} + \binom{k-2}{2}\binom{t+3}{3}\right\}Y^{k+t} \, .$$

If we equate powers of x^n in this relation we obtain formula (12).

4. <u>Concluding remarks.</u> If d_{ij} denotes the distance between
nodes i and j in the tree T, then

$$\sum_{i=1}^{n-1} m_i = \sum_{i=1}^{n-1} d_{in}$$

and

$$\sum_{i=1}^{n} c_i = \sum_{1\le i<j\le n} (d_{ij} - 1).$$

These equations may be established by considering the contribution of
each node or each pair of nodes of T to the sums on the left-hand
sides. If $E(d)$ denotes the expected distance between any two specified nodes over all trees T, then it follows from these equations
that

$$E(m_i) = E(d)$$

and

$$\mu(n) = \frac{1}{2}(n-1)(E(d) - 1).$$

Thus determining $E(m_i)$ and $\mu(n)$ is equivalent to determining $E(d)$; Meir and Moon [4] showed that

$$E(d) = \frac{1}{n-1} \sum_{t=2}^{n} t(t-1) \frac{(n)_t}{(n)^t}.$$

If we compare this formula with formula (4) we obtain another proof of identity (3). We remark that although the first moments $E(m_i)$ and $E(d)$ are equal, the second moments are not equal, since $E(d^2) \sim 2n$.

The first few values of some of the numbers we have been considering are given in the following table.

n	$E(m_i)$	$E(m_i^2)$	$\mu(n)$	$\gamma(n)$
2	1	1	0	0
3	1.3333	2	.3333	.3333
4	1.6250	3.2500	.9375	2.0625
5	1.8880	4.7200	1.7760	6.8160
6	2.1296	6.3889	2.8241	16.6435

References

1. L. E. Clarke, On Cayley's formula for counting trees, J. London Math. Soc. 33 (1958), 471-474.

2. F. Harary, Graph Theory, Addison-Wesley, Reading, 1969.

3. F. Harary and P. Ostrand, The cutting center theorem for trees, Discrete Math. 1 (1971), 7-18.

4. A. Meir and J. W. Moon, The distance between points in random trees, J. Combinatorial Theory 8 (1970), 99-103.

5. J. W. Moon, The expected number of inversions in a random tree, Proceedings of the Louisiana Conference on Combinatorics, Graph Theory and Computing, Louisiana State University, Baton Rouge, 1970, 375-382.

6. J. W. Moon, Counting Labelled Trees, <u>Canad</u>. <u>Math</u>. <u>Cong</u>., Montreal, 1970.

7. J. W. Moon, Random walks on random trees, <u>J</u>. <u>Aust</u>. <u>Math</u>. <u>Soc</u>., to appear.

8. J. Riordan, <u>Combinatorial</u> <u>Identities</u>, Wiley, New York, 1968.

ON THE GIRTH AND GENUS OF A GRAPH

E. A. Nordhaus
Michigan State University
East Lansing, MI 48823

1. Introduction. We present a brief historical discussion of
genus problems in graph theory and some recent results in this area.
An inequality readily obtained from the extended Euler polyhedral
formula is applied to connected regular graphs of degree r and
sufficiently large girth g to provide numerous counterexamples to a
conjecture made by Duke [5] relating the genus and the Betti number of
a graph.

2. Preliminaries. The graphs G considered are connected unless otherwise noted. For graph theory terminology, see Harary [8].

A graph G is said to have genus $k = \gamma(G)$ if it can be imbedded in a compact orientable 2-manifold (surface) of genus k, where
k is minimal. Such surfaces are known to be homeomorphic to spheres
S_k having k handles, or k holes. The word "imbedding" of course
implies that no two edges of G intersect except possibly at a common
end vertex. We will not discuss imbeddings of graphs on non-orientable
surfaces in this paper.

In general, it is not easy to determine the genus of a graph, and
relatively few families of graphs are known for which the genus has
been precisely calculated. These include the complete graphs, the
complete bipartite graphs, and the n-cube.

Recently A. T. White [16, 17, 18] has obtained exact formulas for
the genus of some complete tripartite graphs, for repeated cartesian
products of certain bipartite graphs, and for some finite groups.

In 1968, Ringel and Youngs [14] completed the determination of
the genus of the complete graph K_p, a task which required about 20
years of effort, obtaining

$$\gamma(K_p) = \left\{ \frac{(p-3)(p-4)}{12} \right\}, \quad \text{for} \quad p \geq 3,$$

where {x} denotes the smallest integer not less than x. This

solved the long-standing Heawood map-coloring conjecture. In 1890
Heawood proved that the chromatic $\chi(S_k)$ of a sphere with k handles
satisfies the inequality

$$\chi(S_k) \leq \left\lceil \frac{7 + \sqrt{1+48k}}{2} \right\rceil, \quad \text{where} \quad k > 0.$$

Here [x] denotes the greatest integer less than or equal to x. His
conjecture was that equality holds for k > 0, and this proved to be
correct. It is remarkable that the "simplest" case k = 0 is still
unsolved and is of course the famous four color problem.

A frequently successful method of finding the genus of a connected graph is to establish a lower bound for the genus by using some
form of the Euler polyhedral formula, and then to construct an imbedding of the graph that attains this lower bound. Such constructions
are often difficult to find. We will describe later a technique discussed by Edmonds [6].

Attention has recently been focused on 2-cell (or cellular) imbeddings of a graph G. An imbedding of G on a compact orientable
2-manifold N is called a 2-cell imbedding if every component of the
complement of G in N is homeomorphic to an open unit disk. These
components are called faces of the imbedding.

Youngs [19] has shown that every minimal imbedding of a connected
graph G is necessarily a 2-cell imbedding and has a maximum number
of faces. For a 2-cell imbedding of such a graph G, the extended
Euler equation applies:

$$F(G) - E + V = 2 - 2\gamma(G).$$

Here V and E denote respectively the number of vertices and edges
of G, and F(G) denotes the number of faces in some 2-cell imbedding of G on $S_{\gamma(G)}$. The imbedding need not be minimal.

3. The maximum genus. We now define the maximum genus
$k = \gamma_M(G)$ of a connected graph G to be the largest value of k for
which G has on S_k a 2-cell imbedding. The requirement of a 2-cell
imbedding is essential, since otherwise arbitrarily many handles could
be added to any orientable 2-manifold in which a graph is imbedded.
With this natural extension of the concept of the genus of a graph, it
is seen that the usual genus $\gamma(G)$ might more appropriately be termed

the minimum genus, and is characterized by the fact that the number of faces in the imbedding is a maximum. It is not difficult to show that a connected graph G has a 2-cell imbedding in S_k if and only if $\gamma(G) \leq k \leq \gamma_M(G)$.

In a recent paper by Nordhaus, Stewart, and White [9], numerous results concerning the maximum genus of a graph are developed, and in particular it is shown that the maximum genus of the complete graph K_p is given by the formula

$$\gamma_M(K_p) = \left\lceil \frac{(p-1)(p-2)}{4} \right\rceil.$$

It follows that the complete graph K_p $(p \geq 3)$, has a 2-cell imbedding in S_k if and only if

$$\left\{ \frac{(p-3)(p-4)}{12} \right\} \leq k \leq \left\lceil \frac{(p-1)(p-2)}{4} \right\rceil.$$

If we define the <u>first</u> <u>Betti</u> <u>number</u> of G as $\beta(G) = E - V + 1$, we can write Euler's formula in the form

$$F(G) = 1 + \beta(G) - 2k,$$

where F(G) is the number of faces in some 2-cell imbedding of G on an orientable surface of genus k. Since $F(G) \geq 1$, we obtain an upper bound for the maximum genus when $k = \gamma_M(G)$:

$$\gamma_M(G) \leq [\beta(G)/2].$$

Here equality holds if and only if the imbedding has one or two faces according as $\beta(G)$ is even or odd, respectively. So far most of the success in computing $\gamma_M(G)$ depends on the fact that for the graphs considered, $\gamma_M(G) = [\beta(G)/2]$.

For the complete bipartite graph $K_{m,n}$, where $m,n \geq 2$, a 2-cell imbedding is possible on S_k if and only if

$$\left\{ \frac{(m-2)(n-2)}{4} \right\} \leq k \leq \left\lceil \frac{(m-1)(n-1)}{2} \right\rceil.$$

The lower bound was found by Ringel [12] in 1965 and the upper bound by Ringeisen [11].

The genus of the n-cube Q_n was found by Ringel [13] and independently by Beineke and Harary [2] as

$$\gamma(Q_n) = 1 + (n-4)2^{n-3}, \quad \text{for } n \geq 2.$$

One readily computes that

$$\gamma_M(Q_n) \leq (n-2)2^{n-2}, \quad \text{for } n \geq 2,$$

and we conjecture that equality holds. Equality has been established by J. Zaks.

In a paper by Nordhaus, Ringeisen, Stewart, and White [10], the authors investigate the conditions under which the minimum and maximum genus of a graph G are equal. It is proved that $\gamma(G) = \gamma_M(G)$ only if both vanish, so G is planar. Moreover G must be a special planar graph called a <u>cactus</u>, that is, a connected planar graph in which each block is a K_2 or a simple cycle, and furthermore must have all cycles vertex disjoint. Analogous to the celebrated theorem of Kuratowski which characterizes graphs for which $\gamma(G) = 0$, graphs with $\gamma_M(G) = 0$ can have no subgraph isomorphic to a subdivision of the graphs H_0 and H_1, where H_0 consists of two triangles with a common vertex, and H_1 consists of two triangles with a common edge. See Figure 1.

H_0: 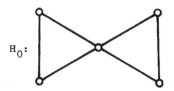 H_1: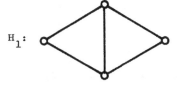

<u>Figure 1.</u>

4. <u>Edmonds' permutation scheme</u>. Two-cell imbeddings can be represented algebraically as follows: let the vertex set of G be $\{1, 2, \ldots, n\}$, and let $V(i) = \{k \mid (i,k) \in E(G)\}$ be the vertices of G adjacent to vertex i, where $E(G)$ is the edge set of G. Let $p_i: V(i) \to V(i)$ be a cyclic permutation of the set $V(i)$, $i = 1, 2, \ldots, n$, of length $n_i = |V(i)|$. It is a theorem of Edmonds [6] that each choice $(p_1, p_2, \ldots, p_n)$ determines a 2-cell imbedding of G in some surface M, such that there is an orientation on M which induces a cyclic ordering of the edges (i,k) at i in which the

immediate successor to (i,k) is (i,p$_i$(k)). Given (p$_1$,p$_2$,...,p$_n$),
there is an algorithm which produces the determined imbedding. Conversely, given a 2-cell imbedding of G in a surface M with a given
orientation, there is a corresponding set of permutations (p$_1$,p$_2$,
...,p$_n$) determining that imbedding.

Let [a,b] and (a,b) denote respectively a directed and an
undirected edge joining vertex a to b. There is an algorithm for
finding all 2-cell imbeddings of a connected graph G. Let
D = {[a,b] | (a,b) $\in$ E(G)}; and define P: D → D by P([a,b]) =
[b,p$_b$(a)]; then P is a permutation on D. The orbits under P
describe the faces of the imbedding. There are

$$\prod_{i=1}^{n} (n_i-1)!$$

possible permutations P. For each of these, one can find the corresponding face distribution.

A. T. White [unpublished] has with the aid of the computer and
the above algorithm found all 7776 2-cell imbeddings of the
Kuratowski graph K$_5$. There are 24 face distributions consistent
with the Euler formula, of which 21 are realizable. The imbeddings
occur only on surfaces S$_1$, S$_2$, and S$_3$.

5. **Duke's conjecture.** We next consider a conjecture made by
R. A. Duke [5], and discussed also in [4] and [7], that for every connected graph G, the following inequality holds between the genus and
Betti number: $\beta(G) \geq 4\gamma(G)$. It is readily verified that this conjecture is valid for all complete graphs and for all complete bipartite
graphs. The conjecture is also obviously true when $\gamma(G) = 0$, and
G is then planar. When G is toroidal, with $\gamma(G) = 1$, then
$\gamma_M(G) > \gamma(G)$, since equality holds if and only if both vanish, and
$\gamma_M(G) \geq 2$. Then $\beta(G)/2 \geq \gamma_M(G)$ implies that $\beta(G) \geq 4\gamma(G)$. It has
recently been proved by M. Milgram [unpublished] that Duke's conjec-
ture is valid when $\gamma(G) = 2$ but that counterexamples arise when
$\gamma(G) \geq 4$. The case $\gamma(G) = 3$ is still undecided.

We will use an inequality easily derived from Euler's extended
formula to construct numerous counterexamples to Duke's conjecture, as
done by P. Ungar [unpublished] using cubic graphs. Let the <u>girth</u>
g(G) of a graph G not a forest be the length of a shortest cycle
in G.

Tutte [15] established the existence of a connected regular graph of given degree (valency) $r \geq 3$ and given girth $g \geq 3$. It follows that there exist such graphs of lowest order. When $r = 3$, a cubic graph of girth g and lowest order exists and is called a g-cage. For a brief discussion of cages, see Behzad and Chartrand [1].

6. __An inequality__. Let G be a connected cubic graph $(r = 3)$ of girth $g \geq 3$, having V vertices and E edges. Then $2E = 3V$ and $\beta(G) = E - V + 1$, so $V = 2(\beta-1)$. In a 2-cell imbedding of G, the boundary of each face has length at least g, although no face need have boundary length g. If F_i denotes the number of faces with i sides, $i \geq g$, then $2E = gF_g + (g+1)F_{g+1} + \ldots \geq gF$, where $F = F_g + F_{g+1} + \ldots$ is the total number of faces. Then

$$F(G) = 1 + \beta - 2\gamma(G) \leq \frac{2E}{g} = \frac{3V}{g} = \frac{6(\beta-1)}{g} ,$$

which implies that

$$\gamma \geq \beta \left(\frac{1}{2} - \frac{3}{g}\right) + \left(\frac{1}{2} + \frac{3}{g}\right).$$

From this inequality

$$\frac{\gamma}{\beta} > \frac{1}{2} - \frac{3}{g} .$$

If $g \geq 12$, then $\beta(G) < 4\gamma(G)$, contradicting Duke's conjecture. It follows that the 12-cage (the Benson graph) [3], or any other connected cubic graph of girth 12 would also provide a counterexample. In general, the genus of the g-cage is not known.

By analogous reasoning, one readily establishes for connected graphs G regular of degree $r \geq 3$ and of girth $g \geq 3$, the inequality

$$\gamma(G) \geq \beta(G)\left(\frac{1}{2} - \frac{\lambda}{g}\right) + \left(\frac{1}{2} + \frac{\lambda}{g}\right),$$

where $\lambda = r/(r-2)$. From this result we obtain counterexamples to Duke's conjecture whenever $g \geq 4\lambda$. An infinite number of such graphs exist.

If G is not a cactus with disjoint cycles, then $\gamma(G) <$ $\gamma_M(G) \leq [\beta(G)/2]$, so that $\beta(G) > 2\gamma(G)$. It can be proved that for $\varepsilon > 0$ there exist r-regular graphs of sufficiently large girth for which $\beta(G) < (2+\varepsilon)\gamma(G)$.

B. M. Stewart has recently proved (unpublished) that there is a unique graph G of valency six and girth six with a minimum number of vertices which has genus 32 and Betti number 125, thus providing another counterexample to Duke's conjecture. The graph G has order 62 and is readily described with the aid of the plane projective geometry PG(2,5) over the field Z_5 of integers reduced modulo 5, and Singer's perfect difference set (0,1,3,8,12,18). This graph is upper embeddable with maximum genus 62.

References

1. M. Behzad and G. Chartrand, Introduction to the Theory of Graphs, Allyn and Bacon, Boston, 1972.

2. L. W. Beineke and F. Harary, The genus of the n-cube, Canad. J. Math. 17 (1965), 494-496.

3. C. T. Benson, Minimal regular graphs of girth eight and twelve, Canad. J. Math. 18 (1966), 1091-1094.

4. R. A. Duke, How is a graph's Betti number related to its genus?, Amer. Math. Monthly 78 (1971), 386.

5. _____, The genus, regional number, and Betti number of a graph, Canad. J. Math. 18 (1966), 817-822.

6. J. R. Edmonds, A combinatorial representation for polyhedral surfaces, Notices Amer. Math. Soc. 7 (1960), 646.

7. R. Guy and V. Klee, Monthly Research Problems, 1969-71, Amer. Math. Monthly 78 (1971), 1113-1122.

8. F. Harary, Graph Theory, Addison-Wesley, Reading, 1969.

9. E. A. Nordhaus, B. M. Stewart, and A. T. White, On the maximum genus of a graph, J. Combinatorial Theory, Ser. B. 11 (1971), 258-267.

10. E. A. Nordhaus, R. D. Ringeisen, B. M. Stewart, and A. T. White, A Kuratowski-type theorem for the maximum genus of a graph, J. Combinatorial Theory, Ser. B. 12 (1972), 260-267.

11. R. D. Ringeisen, Determining all compact orientable 2-manifolds upon which $K_{m,n}$ has 2-cell imbeddings, J. Combinatorial Theory, Ser. B. 12 (1972), 101-104.

12. G. Ringel, Das Geschlecht des vollständigen paaren Graphen, _Abh._ _Math_. _Sem_. _Univ_. _Hamburg_ 28 (1965), 139-150.

13. _____, Über drei Kombinatorische Problem am n-dimensionalen Würfel and Würfelgitter, _Abh_. _Math_. _Sem_. _Univ_. _Hamburg_ 20 (1955), 10-19.

14. G. Ringel and J. W. T. Youngs, Solution of the Heawood Map-coloring problem, _Proc_. _Nat_. _Acad_. _Sci_. _U.S.A_. 60 (1968), 438-445.

15. W. T. Tutte, _Connectivity_ _in_ _Graphs_, Univ. of Toronto Press, 1966.

16. A. T. White, The genus of the complete tripartite graph $K_{mn,n,n}$, _J_. _Combinatorial_ _Theory_ 7 (1969), 283-285.

17. _____, The genus of repeated cartesian products of bipartite graphs, _Trans_. _Amer_. _Math_. _Soc_. 151 (1970), 393-404.

18. _____, On the genus of a group, _Trans_. _Amer_. _Math_. _Soc_., to appear.

19. J. W. T. Youngs, Minimal imbeddings and the genus of a graph, _J_. _Math_. _Mech_. 12 (1963), 303-315.

VARIATIONS OF THE CELL GROWTH PROBLEM*

Edgar M. Palmer
Michigan State University
East Lansing, MI 48823

1. **The cell growth problem.** The "triangular animals" with at
most five cells are displayed in Figure 1. Evidently an animal grows
in the plane by adding a triangular cell of the same size to any of
its sides. Moreover these organisms are assumed to be simply connected. The problem of enumerating triangular animals, square

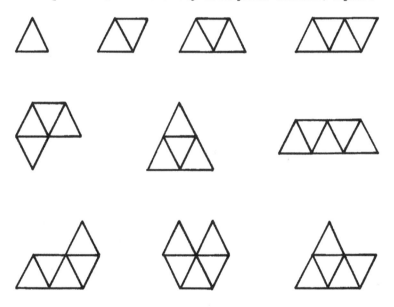

Figure 1. The smallest triangular animals

animals, or hexagonal animals is one of the most difficult in graphical enumeration. One can also ask for the number of toroidal animals,
in which cell growth takes place on a surface of genus 1; or the cells

*   Research supported in part by the National Science Foundation.

can be cubes instead of squares and the problem is to find the number
of solid animals. A complete discussion of this problem, which includes all known data and bounds, is provided in Harary [6]. Definitions not presented in this article can be found in books [7,9].

Special cases of the cell growth problem can be devised if we associate with each animal a graph (called the "characteristic graph" in
[1]) whose points are the animal's cells and in which two points are
adjacent whenever the corresponding cells meet at an edge. The
simplest graphs so obtained are trees. In particular one can then ask
for the number of animals which have as their characteristic graphs a
path of order n.

Thus the generating function which counts square animals whose
characteristic graphs are paths begins

$$x + x^2 + 2x^3 + 3x^4 + 7x^5 + \ldots,$$

and from the data in [1] the series for hexagonal animals begins

$$x + x^2 + 2x^3 + 4x^4 + 10x^5 + 24x^6 + 67x^7 + \ldots \ .$$

The difficulties of the original cell growth problem are still
encountered in this special case because these animals, when unsuper-
vised, can also grow in such a fashion as to wind around and cause
overlapping cells.

2. Variations. The cells of animals are always triangles,
squares or hexagons because these are the only regular polygons which
can fill the plane. But we now show that the problem can be solved
when regular n-gons are used and the organisms are allowed to grow in
3-dimensional space. Present techniques in enumerative analysis are
sufficient only for such extreme modifications.

A class of graphs, denoted Q_n , is defined inductively as follows. The cycle of order n is in Q_n . If G is a graph in Q_n ,
then so is the graph obtained from G by identifying a line of a new
cycle of order n with any line of G. Thus for any (p,q)-graph so
constructed from r cycles of order n, we have

(1) $$p = (n-2)r + 2$$

and

$$q = (n-1)r + 1. \tag{2}$$

The graphs in Q_3 correspond precisely to the 2-dimensional acyclic simplicial complexes called 2-trees. These were enumerated in [8] and the generating function, which has as the coefficient of x^r the number of graphs in Q_3 with exactly r triangles, begins

$$x + x^2 + 2x^3 + 5x^4 + 12x^5 + \dots . \tag{3}$$

The counting series for graphs in Q_4, constructed from cycles of order 4, begins

$$x + x^2 + 3x^3 + 8x^4 + 32x^5 + \dots , \tag{4}$$

(see Figure 2). The procedure of [8] can be modified so that these coefficients can be calculated for any collection Q_n.

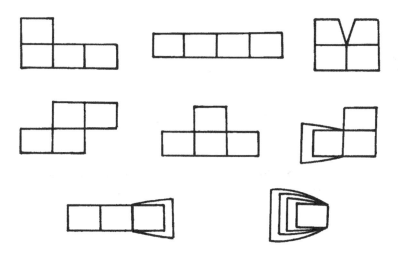

Figure 2. The 8 graphs in Q_4 with 4 cycles of order 4

The labeled counting problem for these graphs is now readily solved and an explicit formula found.

THEOREM 1. <u>The number of ways in which the</u> (p,q)-<u>graphs in</u> $\mathcal{G}_n$, <u>constructed from</u> r <u>cycles of order</u> n, <u>can be labeled is</u>

(5)
$$\frac{p! \, q^{r-2}}{2 \cdot r!}.$$

We note that when $n = 3$, we have $r = p-2$ and $q = 2p-3$ and this formula gives the number of labeled 2-trees [2,3,11].

<u>Proof</u>. This proof employs generating functions for the graphs in $\mathcal{G}_n$ rooted at an oriented line. Functional equations are obtained which can be solved using Lagrange's formula [9; 12, p. 146] and Abel's generalization of the binomial theorem. The desired result follows quickly from the fact that each of these graphs can be rooted at any of its lines.

A <u>rooted graph</u> has one of its lines (the root line) distinguished. A <u>labeled rooted graph</u> has all of its points labeled except those of the root line. A graph in $\mathcal{G}_n$ is <u>simply-rooted</u> if the root line is oriented and it belongs to exactly one cycle of order n.

Let $T_{(n-2)r}$ be the number of labeled, simply rooted graphs of order $p = (n-2)r + 2$ in $\mathcal{G}_n$ and let the exponential generating function $T(x)$ which enumerates these be defined by

(6)
$$T(x) = \sum_{r=1}^{\infty} T_{(n-2)r} \frac{x^{(n-2)r}}{((n-2)r)!}.$$

From the familiar Labeled Counting Lemma (see [9]) which interprets the products of exponential counting series, it follows that

(7)
$$\sum_{k=0}^{\infty} \frac{T(x)^k}{k!}$$

counts the labeled, rooted graphs from $\mathcal{G}_n$ in which the root line is oriented, including the degenerate case where the graph consists only of the oriented root line. On raising the expression (7) to the power $n-1$ we have the series which counts the ways in which these labeled rooted graphs may be attached to the $n-1$ lines of a new cycle of order n to form a simply-rooted graph. Multiplication by x^{n-2} has the effect of installing labels on the $n-2$ points of this new cycle which are not incident with the root edge. Thus we have

$$T(x) = x^{n-2} (\exp T(x))^{n-1}. \tag{8}$$

On setting $z = x^{n-2}$ and $y = T(z^{1/(n-2)})$, formula (8) becomes

$$y = z\, e^{(n-1)y}. \tag{9}$$

From Lagrange's formula we have

$$y = \sum_{r=1}^{\infty} \frac{z^r}{r!} \left\{ \left(\frac{d}{dy}\right)^{r-1} e^{(n-1)ry} \right\}_{y=0} = \sum_{r=1}^{\infty} \frac{((n-1)r)^{r-1}}{r!} z^r. \tag{10}$$

Hence the series for labeled simply-rooted graphs in G_n is:

$$T(x) = \sum_{r=1}^{\infty} \frac{((n-1)r)^{r-1}}{r!} x^{(n-2)r} \tag{11}$$

and

$$T_{(n-2)r} = ((n-2)r)! \frac{((n-1)r)^{r-1}}{r!}. \tag{12}$$

Let $R_{(n-2)r}$ be the number of labeled rooted graphs of order $p = (n-2)r + 2$ in G_n for which the root line is oriented (and permitted to belong to any number of cycles). The generating function for these is

$$R(x) = 1 + \sum_{r=1}^{\infty} R_{(n-2)r} \frac{x^{(n-2)r}}{((n-2)r)!}. \tag{13}$$

Since $T(x)^k/k!$ counts these graphs when the root line belongs to exactly k cycles of order n, it follows that

$$(14) \qquad R(x) = \sum_{k=0}^{\infty} \frac{T(x)^k}{k!} = e^{T(x)} \ .$$

Therefore the coefficients of $R(x)$ and $T(x)$ are related by

$$(15) \qquad \frac{R_{(n-2)r}}{(n-2)r!} = \frac{1}{((n-2)r)} \sum_{k=0}^{r-1} (n-2)(r-k) \frac{R_{(n-2)k}}{((n-2)k)!} \frac{T_{(n-2)(r-k)}}{((n-2)(r-k))!} \ .$$

Equation (11) specifies the values of $T_{(n-2)(r-k)}$ and on substituting these in (15) we have:

$$(16) \qquad \frac{R_{(n-2)r}}{((n-2)r)!} = \frac{1}{r} \sum_{k=0}^{r-1} (r-k) \frac{R_{(n-2)k}}{((n-2)k)!} \frac{((n-1)(r-k))^{r-k-1}}{(r-k)!} \ .$$

To solve this equation explicitly for $R_{(n-2)r}$ we use Abel's generalization of the binomial theorem (see Riordan [13, p.18]), which states:

$$(17) \qquad x^{-1}(x+y+ma)^m = \sum_{k=0}^{m} \binom{m}{k} (x+ka)^{k-1} (y+(m-k)a)^{m-k} \ .$$

On dividing both sides of this identity by $(m+1)!$ and substituting $x = 1$, $y = a = n - 1$ and $m = r - 1$ we have

$$(18) \qquad \frac{((n-1)r+1)^{r-1}}{r!} = \frac{1}{r} \sum_{k=0}^{r-1} (r-k) \frac{((n-1)k+1)^{k-1}}{k!} \frac{((n-1)(r-k))^{r-k-1}}{(r-k)!} \ .$$

Thus (16) and (18) imply

$$(19) \qquad \frac{R_{(n-2)r}}{((n-2)r)!} = \frac{((n-1)r+1)^{r-1}}{r!} \ .$$

The proof is concluded by observing that the coefficient in $R(x)$ of $x^{(n-2)r}$, which is given by the right side of (19), is the sum of the reciprocals of the orders of the automorphism groups of all graphs with $p = (n-2)r + 2$ points in $\mathcal{G}_n$ which are rooted at an oriented

line. On dividing this coefficient by 2 the same sum for graphs
rooted at an unoriented line is obtained. Subsequent multiplication
by $p!/q$ labels <u>all</u> the points and eliminates the root.

Next we enumerate a subclass of labeled graphs from G_n which
bear a closer resemblance to animals. Let H_n denote those graphs in
G_n for which at most two cycles of order n meet at a line. The
graphs in H_3 correspond precisely to the "planar 2-trees" of [8] as
well as triangulations of polygons [5]. The latter can be counted
using the formula of Euler (see [4]) which states that the number of
triangulations of a p-gon which is rooted at an oriented boundary edge
is

$$\frac{1}{p-2}\binom{2(p-2)}{p-3} . \tag{20}$$

In [5], R. K. Guy computed the number of triangulations of p-gons for
$p \leq 25$.

THEOREM 2. <u>The</u> <u>number</u> <u>of</u> <u>ways</u> <u>to</u> <u>label</u> <u>the</u> (p,q)-<u>graphs</u> <u>in</u> H_n,
<u>where</u> <u>at</u> <u>most</u> <u>two</u> <u>of</u> <u>the</u> r <u>cycles</u> <u>of</u> <u>order</u> n <u>can</u> <u>meet</u> <u>at</u> <u>a</u> <u>line</u>,
<u>is</u>

$$\frac{(q-1)!}{2 \cdot r!} . \tag{21}$$

Proof. As in the proof of Theorem 1, $T(x)$ is the exponential
generating function for labeled, simply-rooted graphs in H_n. It is
easy to show that $T(x)$ satisfies

$$T(x) = x^{n-2}(1 + T(x))^{n-1} . \tag{22}$$

On setting $z = x^{n-2}$ and $y = T(x)$, this equation (22) becomes

$$y = z(1+y)^{n-1} . \tag{23}$$

The latter can be solved for y using Lagrange's formula with the result that

$$T(x) = \sum_{r=1}^{\infty} \frac{1}{((n-2)r+1)} \binom{(n-1)r}{(n-2)r} x^{(n-2)r} . \tag{24}$$

Note that for $n = 3$ the coefficient $x^{(n-2)r}$ should be the number of triangulations of a p-gon rooted at an oriented edge. This is quickly verified by substituting $p = r + 2$ in (20). The exponential generating function $R(x)$ for these labeled graphs in $\mathcal{H}_n$ rooted at an oriented line is expressed in terms of $T(x)$ by

$$(25) \qquad R(x) = T(x) + T(x)^2/2 .$$

To obtain an explicit formula for the coefficients of $T(x)^2/2$ we apply Lagrange's formula again to (23):

$$(26) \qquad y^2/2 = \sum_{r=1}^{\infty} \frac{z^r}{r!} \left\{ \left(\frac{d}{dy}\right)^{r-1} y(1+y)^{(n-1)r} \right\}_{y=0} .$$

On carrying out the differentiation and resubstituting for y and z we have

$$(27) \qquad T(x)^2/2 = \sum_{r=2}^{\infty} \frac{1}{r} \binom{r(n-1)}{r-2} x^{(n-2)r} .$$

From the explicit formulas (24) and (27) we can show that the coefficient of $x^{(n-2)r}$ in $R(x)$ is

$$(28) \qquad \frac{R_{(n-2)r}}{((n-2)r)!} = \frac{((n-1)r+1)!}{r!((n-2)r+2)!} .$$

As in Theorem 1 we multiply this number by $p!/2q$ to obtain the desired formula.

Presumably the labeled graphs in $\mathcal{G}_n$ which have either 1 or $m > 2$ cycles of order n meeting at a line can be handled in a similar manner.

The number of plane, labeled graphs of order p from $\mathcal{G}_3$ (i.e. the number of plane labeled 2-trees) was found in [12] to be:

$$(29) \qquad p(p-1)^2 \frac{(5p-10)!}{(4p-6)!} .$$

The corresponding unlabeled problem was also solved. The same approach probably serves to solve both such problems for $\mathcal{G}_n$ with $n > 3$.

In [10], Harary and Read enumerated a subclass of $\aleph_6$ which they called "tree-like polyhexes." Recall that the graphs in $\aleph_6$ are constructed from hexagons and no more than two hexagons are permitted to meet at a line. The graphs in $\aleph_6$ for which no three hexagons are allowed to have a point in common correspond to "tree-like polyhexes." These can also be viewed as certain graphs in $\mathcal{G}_6$ whose "characteristic graphs" are trees (see Figure 3).

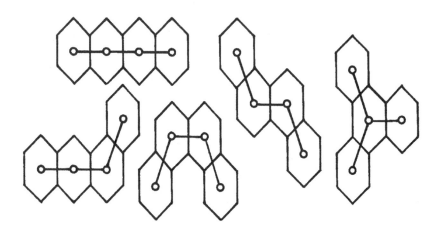

Figure 3. Tree-like polyhexes of four hexagons

This suggests the problem of counting the graphs in $\mathcal{G}_n$ and $\aleph_n$ for which any cycle of order n is permitted to meet other cycles at alternate lines only. The solution would require the formula for the number of ways k people can be seated on a bench of n seats so that no two sit next to each other:

$$\binom{n - k + 1}{k} . \tag{30}$$

All the classes of graphs discussed so far were constructed from cycles of the same length. If cycles of various specified different lengths are used, the enumeration problems are much more complicated but they appear to be manageable.

References

1. A. T. Balaban and F. Harary, Chemical Graphs V, Enumeration and proposed nomenclature of benzenoid cata-condensed polycyclic aromatic hydrocarbons, Tetrahedron 24 (1968), 2505-2516.

2. L. W. Beineke and J. W. Moon, Several proofs of the number of labeled 2-dimensional trees, Proof Techniques in Graph Theory, (F. Harary, Ed.), Academic Press, 1969, 11-20.

3. L. W. Beineke and R. E. Pippert, On the number of k-dimensional trees, J. Combinatorial Theory 6 (1969), 200-205.

4. W. G. Brown, Historical note on a recurrent combinatorial problem, Amer. Math. Monthly 72 (1965), 973-977.

5. R. K. Guy, Dissecting a polygon into triangles, Bull. Malayan Math. Soc. 5 (1968), 57-60.

6. F. Harary, Graphical Enumeration Problems, Graph Theory and Theoretical Physics, Academic Press, London, 1967, 1-41.

7. F. Harary, Graph Theory, Addison-Wesley, Reading, 1969.

8. F. Harary and E. M. Palmer, On acyclic simplicial complexes, Mathematika 15 (1968), 115-122.

9. F. Harary and E. M. Palmer, Graphical Enumeration, Academic Press, to appear.

10. F. Harary and R. C. Read, The enumeration of tree-like polyhexes, Proc. Edinburgh Math. Soc. 17 (1970), 1-13.

11. E. M. Palmer, On the number of labeled 2-trees, J. Combinatorial Theory 6 (1969), 206-207.

12. E. M. Palmer and R. C. Read, On the number of plane 2-trees, J. London Math. Soc., to appear.

13. J. Riordan, Combinatorial Identities, Wiley, New York 1968.

ON THE TOUGHNESS OF A GRAPH

Raymond E. Pippert
Purdue University at Fort Wayne
Fort Wayne, IN 46805
and
Western Michigan University
Kalamazoo, MI 49001

1. __Introduction__. The "toughness" of a graph is an invariant recently introduced by Chvátal [1] who observed some relationships between this parameter and the existence of hamiltonian cycles in the given graph, and also obtained several results regarding this new invariant. In order to enhance the usefulness of toughness as a tool for analyzing graphs, we continue here the investigation of this parameter.

The original approach to toughness is as follows. Let G be a graph and t be a real number such that the implication

$$k(G-S) > 1 \Rightarrow |S| \geq t \cdot k(G-S)$$

holds for each set S of points of G. (We use $k(H)$ to denote the number of components of the graph H.) Then G is said to be __t-tough__. If G is not complete, then there is a largest t such that G is t-tough; this number will be called the __toughness__ of G and will be denoted by $\tau(G)$. Since a complete graph has no cutset S, we set $\tau(K_p) = \infty$ for all p (≥ 1).

An alternate definition is easier to apply in some cases. Let $G(p,q)$ be a graph of connectivity κ, $G \neq K_p$, and let

$$k_n = \max_{|S|=n} \{k(G-S)\} \quad \text{and} \quad t_n = n/k_n. \quad \text{Then} \quad G \text{ is } \underline{t\text{-tough}} \text{ for}$$

$0 \leq t \leq \min_{\kappa \leq n} t_n$. As before, K_p is t-tough for all $t \geq 0$. It follows that if G is not complete, we have $\tau(G) = \min_{\kappa \leq n} t_n$. We also note that there is no need to consider values of n greater than $p - \beta_0$, since t_n is increasing beyond that point.

Before continuing with further results on toughness, we make some preliminary observations. The removal of n points from a cycle results in at most n components, so that $\tau(C_p) \geq 1$, where C_p denotes the cycle with p points. But if $p \geq 4$, it is clear that two components can be obtained by the removal of two points, so that $\tau(C_p) \le 1$ by the definition; hence $\tau(C_p) = 1$. From this, one easily ascertains that every hamiltonian graph is 1-tough. Among numerous conjectures made by Chvátal regarding the implications of toughness on other properties of graphs is the conjecture that $\tau(G) > 3/2$ implies that G is hamiltonian; he also observes that the truth of this conjecture would immediately strengthen the result of Fleischner [2] that the square of every block is hamiltonian. We observe in addition that the toughness of a graph gives some information regarding its structure; for example, the toughness of the complete n-partite graph $K_{m,m,\ldots,m}$ is $n - 1$, regardless of the value of $m \ge 2$.

2. <u>Known results</u>. Among the properties obtained by Chvátal [1] are the following:

A. <u>For any graph</u>, $\tau \ge \kappa/\beta_0$.

B. <u>If</u> τ <u>is finite, then</u> $\tau \le \kappa/2$.

C. <u>If</u> $\beta_0 \ge 2$, <u>then</u> $\tau \le (p-\beta_0)/\beta_0$.

D. <u>If</u> $m \le n$ <u>and</u> $2 \le n$, <u>then</u> $\tau(K_{m,n}) = m/n$.

E. <u>If</u> $m,n \ge 2$, <u>then</u> $\tau(K_m \times K_n) = (m+n-2)/2$.

F. <u>For any graph</u> G, $\tau(G^2) \ge \kappa(G)$.

Properties A and B seem to imply a relationship between toughness and connectivity; however we observe from D that a graph may have high connectivity but low toughness (by appropriate choice of m and n in $K_{m,n}$).

3. <u>Further properties of toughness</u>. The effects of removing (or adding) a point or a line will be considered first. We note that the removal of a point of a graph may raise the toughness arbitrarily. This may be seen by constructing a graph G as follows. Let G' be a graph of toughness n, and form G from G' by adding a point v adjacent to 2k points of G' (we assume $n > k$). Then $\tau(G) \le k$, since the removal of 2k points yields two components, but $\tau(G-v) = \tau(G') = n > k$. Of course, the removal of a point from a graph may result in a complete graph, whereupon the toughness becomes infinite.

On the other hand, the removal of a point cannot lower the toughness greatly.

THEOREM 1. **For any nontrivial graph** G, $\tau(G-v) \geq \tau(G) - 1/2$.

Proof. We let $G' = G - v$ and denote $\tau(G')$ by τ', $\tau(G)$ by τ. If $G' = K_{p-1}$, then $\tau' = \infty$ and the theorem holds. Hence, we assume $G' \neq K_{p-1}$. Let S' be a set of points of G' whose removal yields τ', i.e., $\tau' = |S'|/k(G'-S')$, and let $|S'| = n$, $k(G'-S') = k_n'$. Then $\tau' = n/k_n'$ or $k_n' = n/\tau'$.

Now let $k_{n+1} = \max_{|S|=n+1} k(G-S)$. Clearly $k_{n+1} \geq k_n'$, since one candidate for determining k_{n+1} is the set $S = S' \cup \{v\}$. Thus we have $n/\tau' = k_n' \leq k_{n+1} \leq (n+1)/\tau$, so that $\tau' \geq \tau n/(n+1)$. But $n + 1$ is the cardinality of some separating set of G, which means $n + 1 \geq k_{n+1}\tau \geq 2\tau$. Then $\tau' \geq \tau(2\tau-1)/2\tau = \tau - 1/2$, since $n/(n+1)$ is an increasing function of n.

To see that the lower bound is attained, let $G = 2K_{n-m} + K_m$, the join of K_m with two copies of K_{n-m}, where n is large relative to m. Let v be one of the points of K_m. Then $\tau(G) = m/2$, but $\tau' = \tau(G-v) = (m-1)/2 = \tau - 1/2$.

Next we consider the effect of removing a line of a graph. While the removal of a line cannot possibly raise the toughness, it can lower it by a limited amount.

THEOREM 2. **Let** G **be a graph such that** $1/2 \leq \tau(G) < \infty$, **and let** e **be a line of** G. **Then**

$$\frac{2\tau^2(G) - \tau(G)}{3\tau(G) - 1} \leq \tau(G-e) \leq \tau(G).$$

Proof. Let $G' = G - e$ and denote $\tau(G')$ by τ', $\tau(G)$ by τ, $\max_{|S|=n} k(G'-S)$ by k_n', and $\max_{|S|=n} k(G-S)$ by k_n. Let $\kappa' \leq n_o < n_1 < \cdots < n_\ell$ be those n which yield τ', i.e., $\tau' = n_i/k'_{n_i}$.

The right-hand inequality, $\tau' \leq \tau$, follows immediately from the definition of toughness, so we concern ourselves with the first inequality. If $k'_{n_i} = k_{n_i}$ for all i, then

$$\tau \le \frac{n_i}{k_{n_i}} = \frac{n_i}{k'_{n_i}} = \tau',$$

so that $\tau = \tau'$. Otherwise there is some i for which $k'_{n_i} > k_{n_i}$. Then $k'_{n_i} = k_{n_i} + 1$, since the removal of a line can yield at most one more component.

Now $\tau \le n_i/k_{n_i}$, which we denote by t, so that $k_{n_i} = n_i/t$. Then

$$\tau' = \frac{n_i}{k'_{n_i}} = \frac{n_i}{k_{n_i}+1} = \frac{n_i}{(n_i/t)+1} = \frac{n_i t}{n_i+t}.$$

But $n_i t/(n_i+t)$ is an increasing function of both n_i and t (for n_i and t positive), so that

$$\tau' \ge \frac{n_o \tau}{n_o+\tau} \ge \frac{\kappa'\tau}{\kappa'+\tau}$$

$$\ge \frac{(\kappa-1)\tau}{\kappa-1+\tau} \quad (\tau \ne 0 \Rightarrow \kappa \ge 1)$$

$$\ge \frac{(2\tau-1)\tau}{(2\tau-1)+\tau} \quad (\text{if } \tau \ge 1/2)$$

and the result follows.

Within the last sequence of inequalities are bounds which are in some cases better than that stated in the theorem and which may be useful where further information is available. However for any cycle C_p, we have $\tau(C_p) = 1$, while $\tau' = \tau(C_p-e) = 1/2$, so that the lower bound is achieved.

The toughness of a graph indicates how badly a graph "breaks up" with the removal of a set of points. This can be made more precise as follows.

PROPOSITION 1. Let $G(p,q)$ be a graph with $0 < \tau(G) < \infty$. If S is a set of n points of G, $n \ge \tau$, then $k(G-S) \le [n/\tau]$. Furthermore $k(G-T) \le [p/(\tau+1)]$ for any set T of points of G.

Proof. Let $S \subseteq V(G)$, with $|S| = n \geq \tau$. We must have $k(G-S) \leq [n/\tau]$, for otherwise $k(G-S) \geq [n/\tau] + 1 > n/\tau$, which implies $\tau > n/k(G-S)$, which contradicts the definition of toughness.

Now we also see that for any set $T \subseteq V(G)$, we have $k(G-T) \leq |V(G-T)|$, so that $k(G-T) \leq \min\{[n/\tau], p-n\}$, where $n = |T|$. If $[n/\tau] \leq p - n$, then $k(G-T) \leq [n/\tau] \leq n/\tau$, which implies $n \geq \tau k(G-T)$, so that

$$k(G-T) \leq [n/\tau] \leq p - n \leq p - \tau k(G-T),$$

which in turn yields $k(G-T) \leq p/(\tau+1)$. On the other hand, if $[n/\tau] > p - n$, then $k(G-T) \leq p - n < [n/\tau] \leq n/\tau$, which implies $n > \tau k(G-T)$, so that $k(G-T) \leq p - n < p - \tau k(G-T)$ and hence $k(G-T) \leq p/(\tau+1)$.

The following corollary is immediate.

COROLLARY. Let $G(p,q)$ be a graph with $0 < \tau(G) < \infty$. Then

$$\beta_0(G) \leq \left[\frac{p}{\tau+1}\right].$$

To see that the corollary, and hence the second inequality of Proposition 1, cannot be improved, let G be the star graph $K_{1,p-1}$. We have $\beta_0(G) = p - 1$ and $\tau(G) = 1/(p-1)$, so $[p/(\tau+1)] = p - 1 = \beta_0(G)$.

We next obtain some bounds on the toughness of a graph.

PROPOSITION 2. If G is a graph with $0 < \tau(G) < \infty$, then

$$\frac{1}{\Delta(G)} \leq \tau(G) \leq \frac{\delta(G)}{2}.$$

Proof. The removal of any point of a connected graph G yields at most $\Delta(G)$ components; similarly, the removal of any n points yields at most $n\Delta(G)$ components. Then $\tau = \min_{|S| \geq \kappa} \frac{|S|}{k(G-S)}$ implies

$$\tau \geq \frac{n}{n\Delta(G)} = \frac{1}{\Delta(G)}.$$

That $\tau \leq \delta(G)/2$ follows from the result B.

COROLLARY. If G is connected and regular of degree r, with $\tau(G)$ finite, then $1/r \le \tau(G) \le r/2$.

There is a class of graphs, regular of degree k (odd), whose toughness achieves the lower bound in the corollary (and hence in Proposition 2). The description of these graphs is rather complex and is omitted here. Chvátal points out that $\tau\left(C_p^{k/2}\right) = k/2$ if k is even, which indicates that the upper bound cannot be improved.

THEOREM 3. Let $G = G_1 + G_2$, where $|V(G)| = p$, $|V(G_i)| = p_i$, $\tau(G) = \tau$, and $\tau(G_i) = \tau_i$. Then if $G \ne K_p$, we have

$$\min\left\{\frac{p}{p_1}\tau_1, \frac{p}{p_2}\tau_2\right\} < \tau \le \min\left\{\frac{p-\beta_{10}}{\beta_{10}}, \frac{p-\beta_{20}}{\beta_{20}}\right\},$$

where β_{i0} is the independence number of G_i.

Proof. By the structure of G, it is apparent that the graph cannot be disconnected without removing one of $V(G_1)$ or $V(G_2)$. Having removed the appropriate set, we can then disconnect the graph by disconnecting the remaining G_1 or G_2. Candidates for τ are thus of the form

$$\frac{n_1+p_2}{k(G_1-S_1)} \quad \text{or} \quad \frac{n_2+p_1}{k(G_2-S_2)},$$

where n_1 is the cardinality of some separating set S_1 of G_1, and n_2 is defined analogously. Then

$$\tau = \min\left\{\frac{n_1+p_2}{k(G_1-S_1)}, \frac{n_2+p_1}{k(G_2-S_2)}\right\},$$

where the minimum is taken over all n_1 and n_2 as defined.

Now let S_1 be a separating set for G_1 having cardinality n_1. Then $\tau_1 \le \frac{n_1}{k(G_1-S_1)}$ or $k(G_1-S_1) \le \frac{n_1}{\tau_1}$, so that

$$\frac{n_1+p_2}{k(G_1-S_1)} \ge \frac{n_1+p_2}{(n_1/\tau_1)} = \left(1 + \frac{p_2}{n_1}\right)\tau_1 ;$$

similarly,

$$\frac{n_2 + p_1}{k(G_2 - S_2)} \geq \left(1 + \frac{p_1}{n_2}\right)\tau_2 ,$$

so that

$$\tau \geq \min\left\{\left(1 + \frac{p_2}{n_1}\right)\tau_1, \ \left(1 + \frac{p_1}{n_2}\right)\tau_2\right\} > \min\left\{\frac{p}{p_1}\tau_1, \ \frac{p}{p_2}\tau_2\right\}$$

with strict inequality since we always have $n_1 < p_1$ and $n_2 < p_2$.
    For the other inequality, we observe that two candidates for τ are

$$\frac{(p_1 - \beta_{10}) + p_2}{\beta_{10}} \quad \text{and} \quad \frac{(p_2 - \beta_{20}) + p_1}{\beta_{20}} ,$$

which yield immediately

$$\tau \leq \min\left\{\frac{p - \beta_{10}}{\beta_{10}}, \ \frac{p - \beta_{20}}{\beta_{20}}\right\} .$$

Although the upper bound appears to be a rather imprecise estimate, classes of graphs which achieve the bound are easily constructed. For example, let $G = K_m + C_n$, n even; then $\beta_0(C_n) = n/2$,
while

$$\tau(G) = \frac{m + n/2}{n/2} = \frac{2m + n}{n} ,$$

which is the desired result. In regard to the lower bound, let
$G_1 = K_n - e$ $(n > 2)$ and $G_2 = K_m$ $(m > 2)$. Then the lower bound for
$\tau(G_1 + G_2)$ from the theorem is given by $\tau_1 p / p_1 = ((m + n - 2)/2) - m/n$,
since $\tau_2 = \infty$. But $\tau(G_1 + G_2) = (m + n - 2)/2$, so the lower bound is
arbitrarily close to τ for n sufficiently large as compared to m.

THEOREM 4. <u>Let</u> G <u>be a graph with</u> $0 < \tau(G) < \infty$, <u>and let</u>
$\tau(G) = \tau$, $\lambda(G) = \lambda$ <u>(line-connectivity)</u>. <u>If we denote the line graph</u>
<u>of</u> G <u>by</u> $L(G)$ <u>and</u> $\tau(L(G))$ <u>by</u> τ', <u>then</u> $\max\{\tau, 1/2\} \leq \lambda/2 \leq \tau'$.

Proof. Let S denote a separating set for $L(G)$, with $S| = n$, and let E be those lines of G which correspond to points of S. Then E is a line-separating set of G. We thus have

$$\tau' = \min_{n \geq \kappa(L(G))} \left(\frac{n}{k(L(G)-S)} \right) \geq \min_{n \geq \lambda(G)} \left(\frac{n}{k(G-E)} \right).$$

But this quantity defines a "line-toughness" of G, which has been shown by Chvátal [1] to be precisely $\lambda/2$. Since $\tau > 0$, G is connected, so that $\lambda \geq \kappa \geq 2\tau$ and $\lambda \geq 1$, which yields the desired conclusion.

Since $L(C_p) = C_p$, we have $\tau(L(C_p)) = \tau(C_p)$, which indicates that the lower bound is attained in some cases. There is no upper bound in general; for if we let $G = K_{1,n}$, then $\tau(G) = 1/n$, while $\tau(L(G)) = \tau(K_n) = \infty$.

We conclude with a result on the toughness of a graph and its complement.

THEOREM 5. If a graph $G(p,q)$ is not K_p or $\overline{K}_p$, then

i) $1/(p-1) \leq \tau(G) + \tau(\overline{G}) \leq (p-1)/2$

and

ii) $\tau(G)\tau(\overline{G}) \leq (p-1)^2/16$.

Proof. The inequality ii) is an immediate consequence of i).

To prove i), we observe that at least one of G and $\overline{G}$ is connected. By Proposition 2, any connected graph G has $\tau(G) \geq 1/\Delta(G) \geq 1/(p-1)$. Then clearly $1/(p-1) \leq \tau(G) + \tau(\overline{G})$.

Now $2\tau(G) \leq \kappa(G) \leq \delta(G) \leq 2q/p$, so that $\tau(G) \leq q/p$. Similarly $\tau(\overline{G}) \leq \delta(\overline{G}) \leq 2\left(\binom{p}{2}-q \right)/p$, which implies $\tau(\overline{G}) \leq \left((p-1)/2 \right) - q/p$. Thus $\tau(G) + \tau(\overline{G}) \leq (p-1)/2$.

To achieve the lower bound in i), we let $G = K_{1,p-1}$. Then $\tau(G) = 1/(p-1)$, but $\overline{G}$ is disconnected so $\tau(\overline{G}) = 0$ and the lower bound is attained. With regard to the upper bound, we see that if $G = C_5$, then $\overline{G} = C_5$, and $\tau(G) = \tau(\overline{G}) = 1$, so that $\tau(G) + \tau(\overline{G}) = 2 = (p-1)/2$; also $\tau(G)\tau(\overline{G}) = 1 = (p-1)^2/16$. This indicates that the upper bounds cannot be improved by any constant factor; however, it appears likely that they are not sharp bounds for large values of p.

References

1. V. Chvátal, Tough graphs and hamiltonian circuits, J. Discrete Math., to appear.

2. H. Fleischner, The square of every non-separable graph is hamiltonian, to appear.

ON THE CYCLIC CONNECTIVITY
OF PLANAR GRAPHS

M. D. PLUMMER
Vanderbilt University
Nashville, TN 37235

1. **Introduction.** A connected graph is <u>cyclically k-connected</u> if
it cannot be disconnected into two components each of which contains a
cycle by the removal of fewer than k lines. The concept of cyclic
connectivity as applied to planar graphs dates to the famous incorrect
conjecture of Tait in 1880 [8] which claimed that every planar cubic
graph which was cyclically 3-connected had a Hamiltonian cycle and
thus "proved" the four color conjecture. Histories of Tait's conjecture and the counterexamples to it which have been discovered in the
period 1946 to the present are available in [3] and [7].

One of the most significant reductions of the four color conjecture involves cyclic connectivity and is due to Birkhoff [1,6].

THEOREM 1. (Birkhoff) <u>To prove the four color conjecture true
for all planar graphs it suffices to prove it true for planar cubic
cyclically 5-connected graphs</u>.

The above studies all involve cyclic connectivity of planar
graphs which are regular of degree three. It is easily proven that no
such graph can be more than cyclically 5-connected. Sachs [7] has
shown that if G is not more than 3-connected and regular of degree
4 or of degree 5, then G is at most cyclically 6-connected.

In this paper we wish to drop the regularity restriction and investigate cyclic connectivity for general planar graphs. Any terminology not defined herein may be found in Harary [2] or in Ore [6].

2. **Main Results.** Let G be a finite graph without loops or
multiple lines which is at least 3-connected, i.e., $\kappa(G) \geq 3$. A set
of lines L in G is a <u>cyclic cutset</u> of G if G − L has two components each of which contains a cycle. The minimum cardinality taken
over all cyclic cutsets of G is called the <u>cyclic connectivity</u> of G
and is denoted $c\lambda(G)$. If no such L exists in G, we set $c\lambda(G) = \infty$.

One sees easily, then, that G is cyclically k-connected if and only if $c\lambda(G) \geq k$.

We note first that $c\lambda(G) = \infty$ if and only if G contains no two point-disjoint cycles. This family of graphs has been characterized by Lovász [5].

THEOREM 2. (Lovász) If G has no two point-disjoint cycles and the degree of every point in G is at least 3, then G belongs to one of the following three families of graphs:

I. $V(G) = \{a,b,c,d_1,d_2,\ldots,d_k\}$, $k \geq 1$. Each d_i is joined to each of a, b, and c and in addition any number of the lines ab, bc, ca may be present (subject of course to the restriction that each point of G have degree ≥ 3).

II. G is a "wheel", i.e., G consists of a cycle C and one other point u not on C, but joined to each point of C.

III. $G = K_5$, the complete graph on 5 points.

COROLLARY 2.1. If G is planar and 4-connected, $c\lambda(G) < \infty$.

On the other hand, we now proceed to show that there are 4-connected planar graphs with arbitrarily large cyclic connectivity. Hereafter we denote the degree of a point v by $\rho(v)$.

THEOREM 3. Given any integer $m \geq 4$, there is a planar 4-connected graph G_m with $c\lambda(G_m) = m$.

Proof. For $m = 4$ and 5 let G_m be the graphs shown in Figure 1(a) and (b) respectively.

Now suppose $m \geq 6$. Let G_m be a "double wheel" graph on m points; i.e., G_m consists of a cycle C on $m - 2$ points together with a point u_1 "inside" C joined to every point of C and a point u_2 "outside" C also joined to each point of C.

Note first that any cycle other than C in G_m contains either u_1 or u_2. Hence if L is a cyclic cutset in G_m separating G_m into two components H_m^1 and H_m^2, then u_1 and u_2 lie in different H_m^i. Suppose $u_1 \in H_m^1$ and $u_2 \in H_m^2$. Note also that neither H_m^1 nor H_m^2 contains C. Let the points of $C \cap H_m^1$ be $a_1, \ldots, a_s$ and let the points of $C \cap H_m^2$ be $b_1, \ldots, b_t$. Then $s + t = m - 2$. Finally note that C must contain at least two lines from L, say x_1 and x_2. Then

$$L' = \{u_1 b_1, u_1 b_2, \ldots, u_1 b_t, u_2 a_1, u_2 a_2, \ldots, u_2 a_s, x_1, x_2\}$$

is a subset of L and hence $m = |L'| \leq |L|$. Thus $c\lambda(G_m) \geq m$.

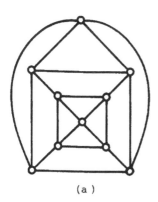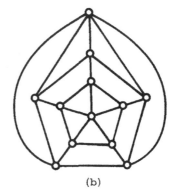

(a) (b)

Figure 1.

It remains to find a cyclic cutset L containing exactly m lines. But if the points of C are $v_1, \ldots, v_{m-2}$, then

$$L = \{u_1 v_1, \ldots, u_1 v_{\frac{m-3}{2}}, u_2 v_{\frac{m-1}{2}}, \ldots, u_2 v_{m-2}, v_1 v_{m-2}, v_{\frac{m-3}{2}} v_{\frac{m-1}{2}}\}$$

is such a set for m odd and

$$L = \{u_1 v_1, \ldots, u_1 v_{\frac{m-2}{2}}, u_2 v_{\frac{m}{2}}, \ldots, u_2 v_{m-2}, v_1 v_{m-2}, v_{\frac{m-2}{2}} v_{\frac{m}{2}}\}$$

is such a set if m is even. Thus $c\lambda(G_m) = m$ and the theorem is proved.

The next theorem points out the interesting fact that if one demands that a planar graph be not only 4-connected, but in fact 5-

connected, the cyclic connectivity becomes bounded above.

THEOREM 4. If G is planar and 5-connected, then $c\lambda(G) \leq 13$.

Proof. We shall appeal to the theory of Euler contributions as developed by Lebesgue [4] and discussed at length in Chapter 4 of [6]. Hereafter, ∂F will denote the boundary of face F.

Let F be a face of G. Define the quantity

$$\phi(F) = 1 - \frac{\rho^*(F)}{2} + \sum_{i=1}^{\rho^*(F)} \frac{1}{\rho_i} ,$$

where $\rho^*(F)$ is the number of lines in ∂F and the ρ_i's are the degrees of the points in ∂F. It may be easily seen [6, pg. 54] that $\Sigma \phi(F) = 2$, where the sum is taken over all faces of G.

Now for a given value of ρ^*, the maximum value of ϕ is given when each $\rho_i = 5$ and we have

$$\phi(F) = 1 - \frac{\rho^*}{2} + \frac{\rho^*}{5} = 1 - \frac{3\rho^*}{10} .$$

Note then that $\phi > 0$ if and only if $\rho^* = 3$, i.e., F is a triangle. If the points of a triangular face F have degrees ρ_1, ρ_2, and ρ_3, respectively, where $\rho_1 \leq \rho_2 \leq \rho_3$, we shall call F a face of type (ρ_1, ρ_2, ρ_3).

Suppose now that $\phi(F) > 0$. Then since

$$\sum_{i=1}^{3} \frac{1}{\rho_i} > \frac{1}{2} ,$$

it may easily be seen [6, pg. 58][1] that F is of type (ρ_1, ρ_2, ρ_3), where either $\rho_1 \leq 5$, $\rho_2 \leq 5$, $\rho_3 \leq 9$, or $\rho_1 \leq 5$, $\rho_2 \leq 6$, and $\rho_3 \leq 7$. Thus the number of lines joining ∂F to the rest of G does not exceed $\max\{5+5+9-6, 5+6+7-6\} = 13$.

---

[1]The reader is advised that the maximal solution types $(4,4,n)$, $(5,5,9)$, and $(5,6,7)$ were inadvertantly omitted on pg. 58 of [6] and should be added to the list of nine types present.

Now let the points of F be v_1, v_2, and v_3. We know $G-v_1-v_2-v_3$ is 2-connected, and since it contains at least three points, it must contain a cycle. Hence $c\lambda(G) \leq 13$ as claimed.

The reader can easily construct planar 5-connected graphs with cyclic connectivity not exceeding 9. In Figure 2 we present an example of a 5-connected planar graph G with $c\lambda(G) = 10$.

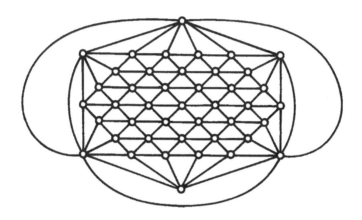

Figure 2.

The fact that this graph G has $c\lambda(G) = 10$ follows immediately from the following theorem.

THEOREM 5. If G is a 5-connected maximal planar graph containing no faces of type $(5,5,5)$, then $c\lambda(G) \geq 10$.

Proof. Let G be a graph satisfying the hypotheses of the theorem. Let L be a cyclic cutset of minimum size in G and let R be one of the components of $G - L$.

Claim. If R has a cutpoint, the theorem follows.

To prove this, first suppose that some block of R is K_2. Call this line x. Let R_1 be a component of $R - x$ which contains a cycle and let R_2 be the other component. Then for $i = 1, 2$ let $L_i \subset L$ be the set of lines joining R_i to $G - R$. Now since G is 5-connected, $|L_i| > 1$, for $i = 1, 2$. But then $L' = (L-L_2) \cup \{x\}$ is a cyclic cutset of lines and $|L'| < |L|$, a contradiction. Thus we may suppose that every block of R contains a cycle. Suppose B_1 and B_2 are two such blocks joined at a cutpoint α. Now blocks B_1 and B_2 are each bounded by cycles C_1 and C_2 respectively of lengths ≥ 3 since G is maximal planar; so let the points of C_i adjacent to α on C_i be β_i and γ_i for $i = 1, 2$. (Cf. Figure 3.)

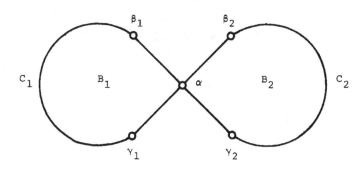

Figure 3.

Now since α is a cutpoint of R, β_1 is not adjacent to β_2 or to γ_2 and β_2 is not adjacent to γ_1. Thus since G is maximal planar, there are at least two lines ℓ_1 and ℓ_2 joining α to $G - R$. Once again, by maximal planarity, there are lines $\ell_3, \ldots, \ell_6$ joining $\beta_1, \beta_2, \gamma_1,$ and γ_2 to $G - R$ respectively.

(i) Suppose B_1 contains a point δ, $\delta \notin V(C_1)$. Then $\{\beta_1, \gamma_1, \alpha\}$ does not separate δ from $G - R$, and in fact there must be at least two more lines ℓ_7 and ℓ_8 joining B_1 to $G - R$.

Similarly, if B_2 contains an interior point, there are two more lines ℓ_9 and ℓ_{10} joining B_2 to $G - R$. So $|L| \geq 10$, as desired.

(ii) Thus suppose one of the B_i, say B_1, contains no interior point. Then $B_1 = C_1$ is a cycle and hence since $\rho(\beta_1), \rho(\gamma_1) \geq 5$, there are at least 6 lines $\ell_1, \ldots, \ell_6$ joining $B_1 - \alpha$ to $G - R$. Thus again $|L| \geq 10$. In either case the claim follows.

Now suppose neither component of $G - L$ (call them R_1 and R_2) has a cutpoint. Thus if F_1 is the face of $G - R_2$ which contains R_2 and if F_2 is the face of $G - R_1$ which contains R_1, both ∂F_1 and ∂F_2 are cycles.

We proceed to build a new graph as follows:

1. Retain only ∂F_1, ∂F_2, and L from G to form H_0.

2. Now ∂F_1 determines a face F_1' of H_0 such that $E(\partial F_1') \cap L = \phi$ and F_2 determines a face F_2' of H_0 such that $E(\partial F_2') \cap L = \phi$. Add two new points u_1 and u_2 to H_0, where u_i is interior to F_1', $i = 1, 2$, and join u_i to each point of $\partial F_1'$, $i = 1, 2$. Call the resulting graph H.

Now H is maximal planar, since G was. Suppose $|L| = \ell$, $|\partial F_1'| = m$, and $|\partial F_2'| = n$. Then

$$\sum \rho(v) = m + m + 2m + \ell + \ell + 2n + n + n = 4m + 4n + 2\ell$$

$$= 2(3|V(G)| - 6) = 2(3(m+n+2) - 6) = 6(m+n).$$

Thus $\ell = m + n$. (This can also be easily proved by induction.) Now if both R_i's contain interior points in G, since $\kappa(G) \geq 5$, it follows that $m, n \geq 5$, and hence that $\ell \geq 10$. So suppose one of the R_i, say R_1, contains no interior point. Then R_1 is a face and hence a triangle. Since G contains no triangle of type $(5,5,5)$ and since $\rho(v) \geq 5$, for all $v \in V(G)$, the sum of the degrees of the points of R_1 must total at least 16 and hence $|L| \geq 10$ and the theorem is proved.

Note that the graph of Figure 2 contains 48 triangular faces, with $\phi > 0$. One can easily show, using the fact that $\Sigma \phi(F) = 2$, that any 5-connected planar graph G with $c\lambda(G) = 10$ must contain at least 30 such faces. Similarly one can also show that any 5-connected planar graph G with $c\lambda(G) = 11, 12,$ or 13 must contain at least 47, 80, or 180 such triangular faces respectively. It is presently unknown whether 5-connected planar graphs with $c\lambda = 11, 12,$ or 13 exist.

Note added in proof: Since this paper was presented examples of 5-connected, cyclically 11-connected, planar graphs have been found independently by J. Malkevitch and B. Grünbaum. The cases for 12 and 13 remain unsettled.

References

1. G. D. Birkhoff, The reducibility of maps, Amer. J. Math. 35 (1913), 115-128.

2. F. Harary, Graph Theory, Addison-Wesley, Reading, Mass., 1969.

3. B. Grünbaum, Polytopes, graphs and complexes, Bull. Amer. Math. Soc. 76 (1970), 1131-1201.

4. H. Lebesgue, Quelques conséquences simples de la formule d'Euler, J. de Math. 9 (1940), 27-43.

5. L. Lovász, Független köröket nem tartalmazó gráfokról, Mat. Lapok 16 (1965), 289-299.

6. O. Ore, The Four-Color Problem, Academic Press, New York, 1967.

7. H. Sachs, Construction of non-Hamiltonian planar regular graphs of degrees 3, 4, and 5 with highest possible connectivity, Theory of Graphs:, International Symposium, Rome, 1966, Gordon and Breach, New York, 1967.

8. P. G. Tait, Remarks on the colouring of maps, Proc. Roy. Soc. Edinburgh 10 (1880), 501-503.

SOME RECENT RESULTS IN CHEMICAL ENUMERATION

Ronald C. Read
University of Waterloo
Waterloo, Ontario, Canada

1. Preliminaries. Before outlining the background and scope of
the problem that we shall consider, it will be as well to review
briefly the basic facts about chemical compounds. For our purposes we
can regard a molecule of a chemical compound as an assemblage of atoms
in which some atoms are linked to others by "valency bonds". These
bonds may be single, double or triple (other kinds will not concern
us). A structural formula is a method of representing the way in
which the atoms in a molecule are linked together. In a structural
formula each atom is represented by a symbol, usually the initial letter or two of its name, e.g., C for carbon, H for hydrogen. A
single bond is represented by a line drawn between the symbols for the
two atoms that it links, a double bond by a double line, and a triple
bond by a three-fold line. Examples of structural formulae are given
in Figure 1. Note that a structural formula makes no attempt to indicate how the atoms are situated in space relative to each other; it
simply indicates which atoms are linked to which, and by what sort of
valency bond. Thus a structural formula, from our point of view, is
essentially a multigraph in which the nodes are of several different
kinds. In this paper we shall be concerned only with hydrocarbons, in
which there are only two kinds of atoms, carbon and hydrogen.

The valency of an atom in a molecule is the number of bonds by
which it is linked to other atoms, double and triple bonds counting as
2 and 3 respectively. Carbon atoms have valency 4, and hydrogen
atoms have valency 1.

An acyclic compound is one for which the structural formula, regarded as a graph, has no cycles, i.e., is a tree. The term "cycle" is
here used in its usual graph-theoretical sense, on the understanding
that double and triple bonds are to be regarded as single edges of the
graph. Thus, although for many purposes a double edge in a multigraph
is regarded as forming a cycle of length 2, we shall not so regard a
double bond. The compounds (a) and (b) in Figure 1 are therefore
acyclic, while (c) is a cyclic compound.

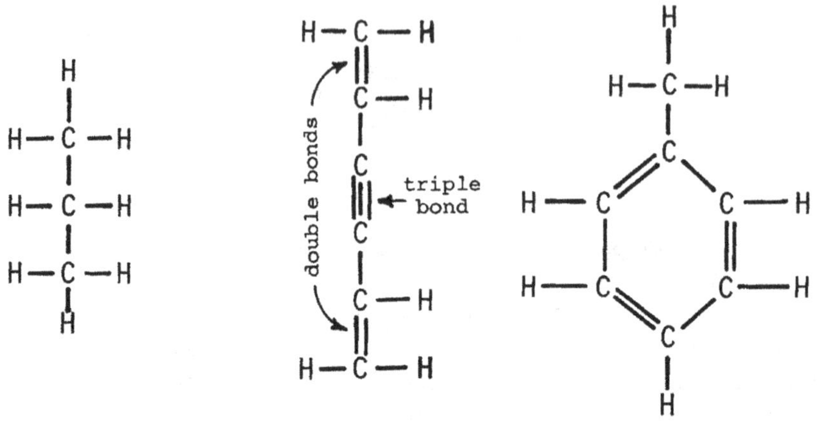

H

|

H—C—H

|

H—C—H

|

H—C—H

|

H

double bonds

H—C—H

‖

C—H

|

C

‖‖ ← triple bond

C

|

C—H

‖

H—C—H

H

|

H—C—H

|

C

H—C   C—H

‖   ‖

H—C   C—H

C

|

H

triple bond

(a). Isopropyl iodide   (b). Hexadienyne-1,5,3   (c). Methyl benzene

Figure 1.

By a <u>radical</u> we shall mean an incomplete molecule -- incomplete
in the sense that there is exactly one valency bond (single, double or
triple) one end of which does not have an atom. Thus a radical is not
a molecule, but something from which molecules can be constructed by
placing an atom, or another radical, at the end of this <u>free</u> bond. If
the free bond is single, double or triple, then the radical is said to
be monovalent, divalent or trivalent respectively. Figure 2(a) shows
the isopropyl radical, having a free single bond. If we put an iodine
atom at the end of the free bond, we get the isopropyl iodide molecule
of Figure 1(a). If instead, we put another isopropyl radical there we
shall get the hydrocarbon of Figure 2(b).

It is easily verified that if an acyclic compound consists entirely of carbon and hydrogen, and has no double or triple bonds, then
its general formula will be C_nH_{2n+2}. That is, if the number of
carbon atoms is n, then the number of hydrogen atoms is 2n + 2.
Such compounds used to be called paraffins, but are now usually called
"alkanes", and the corresponding radicals are called "alkyl radicals".
Thus Figure 2(a) shows an alkyl radical; Figure 2(b) shows an alkane.

The above discussion of nomenclature and notation is greatly
oversimplified, and would not satisfy a professional chemist. However,
it will be sufficient for our needs.

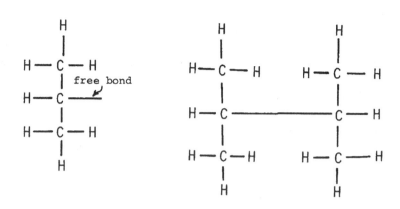

(a). Isopropyl radical   (b). 2,3-dimethyl butane

Figure 2.

2. <u>Background</u>. A problem that has provided a great stimulus to
the development of many tools and results in Combinatorial Analysis is
that of enumerating all chemical compounds with formulae of a given
kind. Problems of this type were studied by Cayley [2, 3] who suc-
ceeded in enumerating the alkyl radicals and the alkanes. Cayley's
results were later extended by other investigators, notably Blair and
Henze [4, 5, 6], and Perry [12]. Enumerative results for other types
of chemical compounds (most acyclic) were obtained, but the methods
whereby this was done remained similar to those used by Cayley. A new
tool for working with problems of this type was forged by Pólya; it is
the "Hauptsatz" of his 1938 paper [13]. The use of this fundamental
theorem enables the findings of Cayley, Blair and Henze, etc. to be
proved much more easily and concisely than by the original methods,
and applications to some kinds of cyclic compounds (benzene derivatives, etc.) became possible.

One somewhat unsatisfactory aspect of the above work was the
transition from the alkyl radicals to the alkanes. It can be seen
from Figure 2(b) that the enumeration of alkanes is essentially the
enumeration of a type of tree; but in an alkyl radical one of the carbon atoms, namely the one incident with the free bond, is different
from the others, so that the enumeration of the alkyl radicals is an

enumeration of <u>rooted</u> trees. The transition in question is therefore essentially that of going from the enumeration of rooted trees to the enumeration of unrooted trees, and at the time of Pólya's paper this could be achieved only by somewhat <u>ad</u> <u>hoc</u> methods. A better method of making this transition was published by Otter [10], and was extended by Harary and Norman [9].

In this paper we shall use all available up-to-date methods in order to enumerate acyclic hydrocarbons with the general formula C_nH_m. We have already remarked that for the alkanes, $m = 2n + 2$. If $m < 2n + 2$, then there must be double or triple bonds present in the molecule. We shall therefore enumerate these compounds according to the number n of carbon atoms that they contain, the number d of double bonds and the number t of triple bonds. Blair and Henze enumerated the ethylene derivatives (having exactly one double bond) and, with Coffman, the acetylene derivatives (having exactly one triple bond), but the problem of enumerating acyclic hydrocarbons with any numbers of double and triple bonds does not seem to have been studied before. Thus this enumeration represents an extension of results already obtained, and the accuracy of this work can be checked, in part, by comparison with these previously known results.

3. <u>Hydrocarbon radicals</u>. In this section we shall enumerate the hydrocarbon radicals, which are of three types, according to the nature of the free bond. The radicals of each of the three types will be classified by the number of carbon atoms, and the numbers of double and triple bonds.

We shall let $G_{n,d,t}$ denote the number of monovalent radicals having n carbon atoms, d double bonds and t triple bonds. Note that we do not need to specify the number of hydrogen atoms, since it is easily verified that this will be $2n + 1 - 2d - 4t$. We similarly define $H_{n,d,t}$ to be the number of divalent radicals (in the sense that the free bond is double) and $I_{n,d,t}$ to be the number of trivalent radicals. These types of radicals are depicted diagrammatically in Figure 3, and will be called, for brevity, G-radicals, H-radicals and I-radicals. In Figure 3 each shaded "balloon" denotes the rest of the radical apart from the free bond and the carbon atom to which it is attached. Note that the free double or triple bond is included in the count for d or t, as the case may be.

G-radical   H-radical   I-radical

Figure 3. The three kinds of radicals

For each of these three types of radical we define a counting series, as follows:

$$G(x,y,z) = \sum_n \sum_d \sum_t G_{n,d,t} x^n y^d z^t \qquad (1)$$

$$H(x,y,z) = \sum_n \sum_d \sum_t H_{n,d,t} x^n y^d z^t \qquad (2)$$

$$I(x,y,z) = \sum_n \sum_d \sum_t I_{n,d,t} x^n y^d z^t \qquad (3)$$

If we examine the G-radicals in closer detail we see that they are of three types, as shown in Figure 4. G-radicals of type (a) can be enumerated by means of Pólya's theorem. We have three positions - the three balloons - each of which can be occupied by a G-radical, or possibly by a hydrogen atom. These three positions can be permuted by any permutation of the symmetric group S_3. Using Pólya's theorem, with figure-counting series $G(x,y,z)$ and group S_3, we obtain the configuration counting series

$$x \ Z(S_3; \ G(x,y,z) \qquad (4)$$

in the notation of Pólya's paper; or rather, we would if the possibility of a G-radical being a single hydrogen atom were included in $G(x,y,z)$. It is therefore convenient to adopt this convention, and regard the radical "H-" as a G-radical, with $n = 0$. Expression (4) is then valid. Note the multiplier "x" in (4), which is needed to account for the carbon atom to which the three radicals are attached.

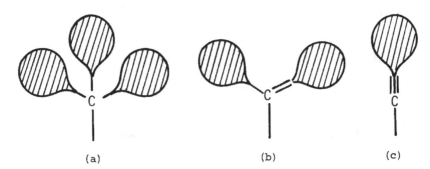

Figure 4. Types of G-radicals

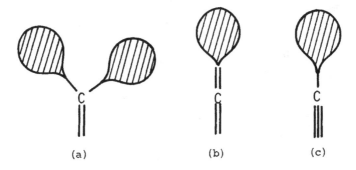

Figure 5. Types of H- and I-radicals

In Figure 4(b) we have two dissimilar positions, into one of which a G-radical must be placed, while an H-radical must be placed in the other. Hence radicals of this type are enumerated by

(5) $$x \ G(x,y,z) \cdot H(x,y,z).$$

In Figure 4(c) there is only one position, in which must be placed an I-radical. Hence these radicals are enumerated by

(6) $$x \ I(x,y,z).$$

Combining these results, and adding a term "1" for the hydrogen radical, we obtain the following equation.

$$G(x,y,z) = 1 + x[Z(S_3;G(x,y,z)) + G(x,y,z)H(x,y,z) + I(x,y,z)] \qquad (7)$$

Looking more closely at H-radicals we see two types, as in Figure 5(a) and 5(b). In Figure 5(a) we have two interchangeable positions in which to place a G-radical. Pólya's theorem then gives

$$xy \ Z(S_2; \ G(x,y,z)) \qquad (8)$$

as the counting series for these, the factor xy being included to account for the free double bond and the incident carbon atom. In Figure 5(b) we have a single position in which to put an H-radical. This gives us the counting series $xy \ H(x,y,z)$. From this, and (8), we derive

$$H(x,y,z) = xy \ Z(S_2; \ G(x,y,z)) + xy \ H(x,y,z) \qquad (9)$$

or

$$H(x,y,z) = \frac{xy}{1 - xy} \ Z(S_2; \ G(x,y,z)) \qquad (10)$$

I-radicals are of one type only (Figure 5(c)), and hence are given by the counting series $xz \ G(x,y,z)$. Thus

$$I(x,y,z) = xz \ G(x,y,z). \qquad (11)$$

We now substitute for H and I, from (10) and (11) in equation (7), and obtain

$$G(x,y,z) = 1 + x[Z(S_3;G(x,y,z)) \qquad (12)$$

$$+ \frac{xy}{1 - xy} \ G(x,y,z) \cdot Z(S_2;G(x,y,z)) + xz \ G(x,y,z)].$$

Equation (12) enables $G(x,y,z)$ to be calculated recursively up to terms in a given power of x. The calculation is extremely tedious by hand, but not too difficult on a computer. Results for $n \leq 10$ are given in Table 1.

As a check on these results, we note that if we put $y = z = 0$ we eliminate all radicals having double or triple bonds, and are left with the alkyl radicals alone. If we write $r(x)$ for $G(x,0,0)$ and perform the indicated substitution in $Z(S_3)$, equation (12) becomes

n=2	t 0	1
0	1	1
d 1	1	

n=3	t 0	1
0	2	2
d 1	3	
2	1	

n=4	t 0	1	2
0	4	4	1
1	8	3	
d 2	5		
3	1		

n=5	t 0	1	2
0	8	10	4
1	21	14	
d 2	20	3	
3	7		
4	1		

n=6	t 0	1	2	3
0	17	25	12	1
1	56	50	7	
2	69	25		
d 3	37	3		
4	9			
5	1			

n=7	t 0	1	2	3
0	39	64	38	7
1	149	166	45	
2	228	134	7	
d 3	165	36		
4	60	3		
5	11			
6	1			

n=8	t 0	1	2	3	4
0	89	166	115	29	1
1	398	531	206	13	
2	725	587	84		
d 3	664	261	7		
4	326	47			
5	88	3			
6	13				
7	1				

n=9	0	t 1	2	3	4
0	211	437	348	114	11
1	1068	1656	829	115	
2	2261	2325	577	13	
3	2505	1470	123		
d 4	1570	433	7		
5	570	58			
6	122	3			
7	15				
8	1				

n=10	0	1	t 2	3	4	5
0	507	1157	1040	417	62	1
1	2876	5076	3103	661	22	
2	6932	8639	3066	222		
3	9032	7121	1155	13		
d 4	6909	2998	162			
5	3204	648	7			
6	915	69				
7	161	3				
8	17					
9	1					

Table I. Monovalent acyclic radicals

$$r(x) = 1 + \frac{1}{6} x[r^3(x) + 3r(x)r(x^2) + 2r(x^3)] \qquad (13)$$

which is the functional equation given by Pólya for the alkyl radicals
(see [13], page 150).

4. <u>Hydrocarbons</u> <u>with</u> <u>a</u> <u>distinguished</u> <u>carbon</u> <u>atom</u>. As stepping-
stones toward our goal of enumerating hydrocarbons, we need two sub-
sidiary results. The first of these is the enumeration of hydrocarbons in which one carbon atom has been distinguished from the others.
We shall denote this distinguished carbon atom by C*.

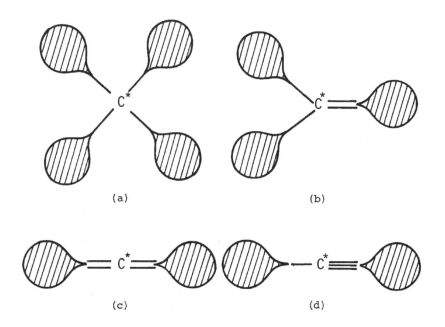

(a)   (b)

(c)   (d)

<u>Figure</u> 6.   Types of hydrocarbon with a distinguished carbon atom

We readily see that these compounds are of four types, shown in
Figure 6. Each of these types can be enumerated using Pólya's theorem.
We find that those in Figure 6(a) are enumerated by

$$x\, Z(S_4;\ G(x,y,z)); \qquad (14)$$

those in Figure 6(b) by

(15)   x $Z(S_2; G(x,y,z))$ $H(x,y,z)$;

those in Figure 6(c) by

(16)   x $Z(S_2; H(x,y,z))$;

and those in Figure 6(d) by

(17)   x $G(x,y,z) \cdot I(x,y,z)$.

Adding these four results we obtain the counting series for the hydrocarbons with a distinguished carbon atom. We shall denote it by $K(x,y,z)$. Hence we have

$$K(x,y,z) = x\, Z(S_4;G(x,y,z)) + x\, Z(S_2;G(x,y,z)) \cdot H(x,y,z)$$
(18)
$$+ x\, Z(S_2;H(x,y,z)) + x\, G(x,y,z) \cdot I(x,y,z)$$

from which $K(x,y,z)$ can be computed, since everything else is known.

5. <u>Hydrocarbons with a distinguished carbon-carbon bond</u>. Our second subsidiary result is the enumeration of hydrocarbons in which one particular carbon-carbon bond has been distinguished from the rest. If this bond is single, then our compound looks like Figure 7, and is clearly the result of joining two monovalent radicals. Since the distinguished bond is to join two carbon atoms, neither of these radicals can be just a hydrogen atom. We therefore have a Pólya type enumeration problem, with two interchangeable positions, into each of which must be put a radical enumerated by $G_1(x,y,z) = G(x,y,z) - 1$. These hydrocarbons are therefore enumerated by

(19) $Z(S_2;G_1(x,y,z)) = \dfrac{1}{2}\left[G_1^{\,2}(x,y,z) + G_1(x^2,y^2,z^2) \right]$

For reasons that will become apparent, we shall be more interested in those hydrocarbons of this kind for which the two radicals are <u>dissimilar</u>. The counting series for these can be found either by subtracting from (19) the counting series for hydrocarbons for which these radicals are identical, or by using a modification of Pólya's theorem (for which see [13], page 172). In either case the counting series turns out to be

$$\frac{1}{2}\left[G_1^{\ 2}(x,y,z) - G_1(x^2,y^2,z^2)\right]. \tag{20}$$

If the distinguished bond is double, then a similar argument gives us

$$\frac{1}{2y}\left[H^2(x,y,z) - H(x^2,y^2,z^2)\right] \tag{21}$$

for those compounds for which the two radicals are dissimilar. The corresponding result for a triple bond is

$$\frac{1}{2z}\left[I^2(x,y,z) - I(x^2,y^2,z^2)\right]. \tag{22}$$

The division by y and z in (21) and (22) results from the fact that in putting the two radicals together we convert what were previously <u>two</u> bonds into one.

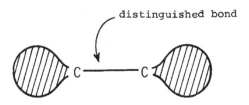

distinguished bond

<u>Figure</u> 7.

6. <u>Acyclic hydrocarbons</u>. We shall now tie together these apparently disconnected results, and obtain the counting series for general acyclic hydrocarbons that we are seeking. To do this we need a theorem, given in [9], which we now briefly describe.

Let T be a tree, and u and v two of its nodes. If there is an automorphism of T which maps u onto v we say that u and v are equivalent. This relation is an equivalence relation, and the nodes of T are therefore partitioned into equivalence classes. Let p* be the number of these classes.

An automorphism of T induces, in an obvious way, a mapping of the set of edges of T onto itself, and we can define equivalence between edges a and b if a maps onto b in some such mapping. The edge set of T is thus partitioned into a number, q* say, of equivalence classes. The theorem is then that

(23) $p^* - (q^* - s) = 1,$

where s = 1 if T has a symmetric edge, and s = 0 otherwise. An edge uv is symmetric if there is an automorphism which maps u onto v and v onto u. Clearly the "half-trees" at the ends of a symmetric edge must be identical, and a tree can have at most one symmetric edge.

If we sum (23) over all trees with a given number p of nodes we obtain

(24 $$\sum p^* - \left(\sum q^* - \sum s \right) = \sum 1 .$$

Now p* is the number of distinct ways (to within automorphism) of choosing a node of T, and is therefore the number of trees with a distinguished node that can be obtained from T. Thus $\Sigma\, p^*$ is the total number of distinct trees rooted at a node. By similar reasoning $\Sigma\, q^*$ is the total number of distinct trees having one edge distinguished from the others.

The term $\Sigma\, s$ in (24) is the number of trees having a symmetric edge, and the term $\Sigma\, 1$ on the right-hand side is simply the total number of trees. Hence this latter number can be computed if $\Sigma\, p^*$, $\Sigma\, q^*$ and $\Sigma\, s$ are known.

For simplicity this argument has been given in terms of ordinary graph-theoretical trees, but it holds just as well for the hydrocarbons that we have been discussing. For any given n, d and t the term $\Sigma\, p^*$ gives the number of hydrocarbons with a distinguished carbon atom, and this is the coefficient of $x^n y^d z^t$ in K(x,y,z) of equation (19). The term $\Sigma\, q^* - \Sigma\, s$ will be the number of hydrocarbons with a distinguished carbon-carbon bond minus the number of these with a symmetric bond. We have anticipated this subtraction in finding (20), (21), and (22). The sum of these three expressions will be a counting series in which the coefficient of $x^n y^d z^t$ is $\Sigma\, q^* - \Sigma\, s$.

Finally, let us denote by $L(x,y,z)$ the counting series in which the coefficient of $x^n y^d z^t$ is the number of hydrocarbons that we are seeking, having n carbon atoms, d double bonds, and t triple bonds. This coefficient is thus the $\Sigma\,1$ of (24). Putting together the results of this section we have

$$L(x,y,z) = K(x,y,z) - \frac{1}{2}\left[G_1^{\,2}(x,y,z) - G_1(x^2,y^2,z^2) \right] \qquad (25)$$

$$- \frac{1}{2y}\left[H^2(x,y,z) - H(x^2,y^2,z^2) \right]$$

$$- \frac{1}{2z}\left[I^2(x,y,z) - I(x^2,y^2,z^2) \right].$$

The coefficients in $L(x,y,z)$ have been calculated for $n \le 10$ and are exhibited in Table II.

7. Deductions and verification. It is of interest to see briefly how (25) links up with previously known results. The coefficient of y in $L(x,y,0)$ will enumerate compounds having just one double bond, i.e. the alkyl derivatives of ethylene. From Table II we extract the information that it is

$$x^2 + x^3 + 3x^4 + 5x^5 + 13x^6 + 27x^7 + 66x^8 + 153x^9 + 377x^{10} + \ldots . \qquad (26)$$

The coefficients in (26) agree with the numbers of ethylene derivatives as given by Blair and Henze [6], who computed them by a completely different method.

Similarly the coefficient of z in $L(x,0,z)$ will enumerate the alkyl derivatives of acetylene (having just one triple bond). From Table II we see that this is

$$x^2 + x^3 + 2x^4 + 3x^5 + 7x^6 + 14x^7 + 32x^8 + 72x^9 + 171x^{10} + \ldots, \qquad (27)$$

in which the coefficients agree with the computations of Coffman, Blair and Henze [8].

The term independent of y and z in $L(x,y,z)$ will enumerate the alkanes. It is

n=2

d	t=0	1
0	1	1
d 1	1	

n=3

d	t=0	1
0	1	1
d 1	1	
2	1	

n=4

d	t=0	1	2
0	2	2	1
1	3	1	
d 2	2		
3	1		

n=5

d	t=0	1	2
0	3	3	2
1	5	4	
d 2	6	1	
3	2		
4	1		

n=6

d	t=0	1	2	3
0	5	7	5	1
1	13	12	3	
2	16	7		
d 3	10	1		
4	3			
5	1			

n=7

d	t=0	1	2	3
0	9	14	11	3
1	27	34	12	
2	44	29	3	
d 3	32	9		
4	15	1		
5	3			
6	1			

n=8

d	t=0	1	2	3	4
0	18	32	28	10	1
1	66	95	48	4	
2	120	110	22		
d 3	115	53	3		
4	62	12			
5	21	1			
6	4				
7	1				

n=9

d	t=0	1	2	3	4
0	35	72	69	28	5
1	153	262	157	29	
2	328	376	120	4	
3	367	354	29		
d 4	253	85	3		
5	100	14			
6	28	1			
7	4				
8	1				

n=10

d	t=0	1	2	3	4	5
0	75	171	179	88	20	1
1	377	718	518	138	8	
2	901	1245	537	53		
3	1196	1074	226	4		
d 4	964	498	39			
5	491	124	3			
6	160	17				
7	36	1				
8	5					
9	1					

Table II. Acyclic hydrocarbons

$$x + x^2 + x^3 + 2x^4 + 3x^5 + 5x^6 + 9x^7 + 18x^8 + 35x^9 + 75x^{10} + \ldots, (28)$$

in which the coefficients agree with those given in [5].

Another interesting result, not previously obtained, is that for the number of hydrocarbons with a given number n of carbon atoms, irrespective of the number of double and triple bonds. This number is the coefficient of x^n in $L(x,1,1)$ and, from Table II, we have

$$L(x,1,1) = x + 3x^2 + 4x^3 + 12x^4 + 27x^5 + 84x^6 + 247x^7 + 826x^8 \qquad (29)$$

$$+ 2777x^9 + 9868x^{10} + \ldots \; .$$

8. <u>Prospects</u>. One naturally asks whether results such as these could be extended further, or generalized. One could consider, for example, removing the restriction to acyclic compounds, but if one does this one is beset by a host of difficulties. The problem is then essentially that of enumerating graphs or multigraphs with a given partition, and although methods have been published for doing this (see [7], [11]) they are all extremely cumbersome and, in general, quite impracticable. The prospect in this direction is not very bright, though some special cases may be amenable to treatment.

One can, instead, retain the restriction to acyclic compounds but allow atoms other than carbon or hydrogen. Oxygen is the obvious next candidate, and I have, in fact, carried out an enumeration, similar to that given here, for acyclic compounds of carbon, oxygen and hydrogen, having given numbers of these atoms, and given numbers of single, double and triple bonds. The results are complicated, tedious and uninspiring, and, what is more, they suffer from a grave defect.

The molecules enumerated by $L(x,y,z)$ are all more or less chemically feasible, and we can therefore cherish the illusion that what we have done in this paper has at least a tenuous connection with real life. When we include oxygen atoms, all such illusions are shattered. For example, chains of any number of oxygen atoms, of the form

$$\ldots - 0 - 0 - 0 - 0 - \ldots .$$

are theoretically possible, whereas chains of more than three oxygen atoms are not chemically feasible. This difficulty can be overcome (at the expense of making the results even messier) but there are other snags. Take, for instance, the hydroxyl radical " -O-H ".

There is no obvious reason why there should not be two such radicals attached to one and the same carbon atom; yet such a configuration is not chemically possible. Considerations such as these make it virtually impossible to derive a series which will count only those carbon-oxygen-hydrogen compounds that are chemically significant; yet any enumeration that includes impossible compounds is pretty much a waste of time. There seems to be little scope in this direction either.

One related problem that is not entirely pointless is that of enumerating molecules taking into account something of the way that they are situated in space. If this is done to the extent of recognizing differences between left-handedness and right-handedness, we have the problem of enumerating stereo-isomers. This problem was discussed by Pólya for the alkanes, and I have extended it to general hydrocarbons, much as in this paper. The results of this enumeration, and others, are to be included in a chapter of a book [1] at present in course of preparation.

References

1. A. T. Balaban, Chemical Applications of Graph Theory, Academic Press, to appear.

2. A. Cayley, On the mathematical theory of isomers, Phil. Mag. 47 (1874), 444-446.

3. _____, On the analytical forms called trees, with applications to the theory of chemical combinations, Report of the British Association for the Advancement of Science, (1875), 257-305.

4. C. M. Blair and H. R. Henze, The number of structurally isomeric alcohols of the methanol series, J. Amer. Chem. Soc. 53 (1931), 3042-3046.

5. _____, The number of isomeric hydrocarbons of the methane series, J. Amer. Chem. Soc. 53 (1931), 3077-3085.

6. _____, The number of structurally isomeric hydrocarbons of the ethylene series, J. Amer. Chem. Soc. 55 (1933), 680-686.

7. C. C. Cadogan, On multigraphs with a given partition, Bull. Australian Math. Soc. 3 (1970), 125-137.

8. D. D. Coffman, C. M. Blair, and H. R. Henze, The number of structurally isomeric hydrocarbons of the acetylene series, J. Amer. Chem. Soc. 55 (1933), 252-253.

9. F. Harary and R. Norman, Dissimilarity characteristic theorems for graphs, Proc. Amer. Math. Soc. 11 (1960), 332-334.

10. R. Otter, The number of trees, Ann. Math. 49 (1948), 583-599.

11. K. Parthasarathy, Enumeration of graphs with given partition, Canad. J. Math. 20 (1968), 40-47.

12. D. Perry, The number of structural isomers of certain homologs of methane and methanol, J. Amer. Chem. Soc. 54 (1932), 2918-2920.

13. G. Pólya, Kombinatorische Anzahlbestimmungen für Gruppen, Graphen und chemische Verbindungen, Acta Math. 68 (1938), 145-254.

UPPER AND LOWER IMBEDDABLE GRAPHS

Richard D. Ringeisen
Colgate University
Hamilton, NY 13346

1. **Introduction.** In this paper we further explore the relationship between genus and maximum genus. In particular we examine those
graphs whose genus attains certain lower bounds and whose maximum
genus attains an upper bound. We also explore the possible relationship between upper or lower imbeddable graphs and the demised conjecture of R. Duke [3]. An exploration of upper imbeddable blocks and of
the effect of such blocks on a supergraph is then initiated.

The <u>Betti number</u>, $\beta(G)$, of a connected graph G with E edges
and V vertices is $E - V + 1$. The genus of a graph is determined by
examining its imbedding in compact orientable 2-manifolds (surfaces).
Such a manifold may be thought of as either a sphere or a sphere with
"handles" attached. The number of handles on a surface N is called
the genus of the surface and is denoted by $\gamma(N)$. By imbedding a
graph on a surface, we mean placing the vertices and edges of the
graph on the surface so that the edges do not cross. A <u>cellular imbed-</u>
<u>ding</u> of a graph H is one in which each of the components of the
complement of H in the surface is homeomorphic to an open disc.

The <u>genus</u>, $\gamma(G)$, of a graph G is the smallest of the numbers
$\gamma(N)$, where N is a compact orientable 2-manifold in which G has a
celluar imbedding. The <u>maximum genus</u> of a connected graph G, $\gamma_M(G)$,
is the largest of the numbers $\gamma(N)$. That such a maximum exists is
insured by the following upper bound theorem [5].

THEOREM. <u>An upper bound for the maximum genus of an arbitrary</u>
<u>connected graph is given by</u> $\gamma_M(G) \leq [\beta(G)/2]$ <u>Equality holds if and</u>
<u>only if the imbedding has one or two faces, according as</u> $\beta(G)$ <u>is</u>
<u>even or odd, respectively.</u>

A connected graph G is <u>upper imbeddable</u> if $\gamma_M(G) = [\beta(G)/2]$.
Equivalently, a connected graph is upper imbeddable if there exists a
surface upon which it has an imbedding with one or two faces.

The complete graphs [5], the complete bipartite graphs [7], the
wheel graphs, the standard maximal planar graphs, and the prism graphs

[8] are all upper imbeddable.

To simplify notation we define the following parameters for a connected graph G:

$$X_1(G) = E/6 - (V-2)/2,$$

$$X_2(G) = E/4 - (V-2)/2.$$

We let $\{y\}$ be the smallest integer larger than or equal to y. If a graph G has the complete graph K_3 as a subgraph, we say G has a triangle. If we exclude the graph K_2, we have the following inequalities [2].

$\gamma(G) \geq \{X_1(G)\}$ if G is any connected graph,

$\gamma(G) \geq \{X_2(G)\}$ if G is a connected graph without triangles.

A connected graph G is lower imbeddable if

i) G has triangles and $\gamma(G) = \{X_1(G)\}$ when $X_1(G) > 0$ or $\gamma(G) = 0$ when $X_1(G) \leq 0$, and

ii) G has no triangles and $\gamma(G) = \{X_2(G)\}$ when $X_2(G) > 0$ or $\gamma(G) = 0$ when $X_2(G) \leq 0$.

The definition implies that any connected planar graph is lower imbeddable. The genus theorems for K_m [9] and $K_{m,n}$ [10] show these graphs are lower imbeddable.

2. A note on Duke's conjecture. In [2] Duke conjectures that for a connected graph G the Betti number $\beta(G) \geq 4\gamma(G)$ (see also [4]). Recently, however, both Martin Milgram and Peter Ungar have produced counterexamples (both results are unpublished). Ungar shows that any connected cubic graph of girth at least twelve gives a counterexample. The conjecture however is known to hold for all planar and toroidal graphs [3].

We show the conjecture is true for all lower imbeddable graphs. The proof involves only the judicious use of simple inequalities.

THEOREM 1. If G is any connected lower imbeddable graph, then $\beta(G) \geq 4\gamma(G)$.

Proof. Since the theorem holds for all toroidal graphs, we assume $\gamma(G) \geq 2$. We are thus considering graphs with eight or more
vertices. Hence, $E + 3V \geq 31$, since G is connected, and thus
$E \geq V - 1$.

Case 1: Suppose G has triangles. Thus $\gamma(G) = \{X_1(G)\}$ and
$\gamma(G) < X_1(G) + 1$. So $4\gamma(G) < 2E/3 - 2V + 8$, which is less than or
equal to $\beta(G)$ if $E + 3V \geq 21$.

Case 2. Suppose G has no triangles. Then, $\gamma(G) = \{X_2(G)\}$
and $\gamma(G) < X_2(G) + 1$. Hence, $4\gamma(G) < E - 2V + 8$, which is less
than or equal to $\beta(G)$ if $V \geq 7$.

The above theorem encourages one to look toward upper imbeddable
graphs for a similar result. If any of the aforementioned counterexamples are upper imbeddable all such thoughts are banished. However,
the author has been unable to obtain such a result. So, we have the
question: Do all upper imbeddable connected graphs satisfy Duke's
conjecture?

3. Imbeddability classifications of graphs. Considerations such
as Duke's conjecture lead us to discuss graphs with regard to comparison of genus and maximum genus. Before we proceed with such a discussion, however, we need the following theorem [6]. It is worth
noting that the maximum genus of a graph need not be the sum of the
maximum genera of its blocks.

THEOREM. Given a graph H with n components $G_1, \ldots, G_n$,
let G be a connected graph obtained from H by the addition of $n-1$
edges. Then

$$\gamma_M(G) = \sum_{i=1}^{n} \gamma_M(G_i).$$

Any connected graph belongs to exactly one of the following four
sets:

A. Graphs which are both upper and lower imbeddable.
B. Graphs which are upper, but not lower imbeddable.
C. Graphs which are lower, but not upper imbeddable.
D. Graphs which are neither upper nor lower imbeddable.

THEOREM 2. Each of the sets described above is an infinite set of graphs.

Proof. We recall from [5] that $\gamma_M(K_n) = [(n-1)(n-2)/4]$, while from [10] $\gamma(K_n) = \{(n-3)(n-4)/12\}$. We proceed by displaying at least one graph in each set. We then note that a generalization of examples will complete the proof.

Set A. As mentioned above, the complete graphs, the complete bipartite graphs, the wheel graphs, and the standard maximal planar graphs provide examples.

Set B. The Petersen graph P is shown to be upper imbeddable later in this paper; while $X_2(P)$ is negative, $\gamma(P)$ is one.

A graph with X_2 positive is provided in the following manner. To a graph with the three components K_3, K_{10}, and K_{10}, we add two edges, each incident with K_3, in such a manner that the resultant graph G is connected. Then the well known genus summability theorem [1], and the maximum genus theorem quoted above give the following information: $\gamma(G) = 8 \neq 6 = \{X_1(G)\}$, $\gamma_M(G) = 36 = [\beta(G)/2]$.

Set C. In [6] it is shown that any cactus with disjoint cycles has maximum genus zero. Any such graph with at least three cycles has Betti number larger than three. Since all such cacti are planar, we have an infinite number of examples.

A nonplanar example is provided by considering a graph with two components, both of which are K_7. If one forms a connected graph G by the addition of a single edge he has an example. Since $X_1(G) = 7/6$ and $\gamma(G) = 2$, G is lower imbeddable. Because $\gamma_M(G) = 14$ while $\beta(G) = 30$, G is not upper imbeddable.

Set D. We give two examples, one with X_1 negative, the other with X_1 positive.

Consider a graph with three components, K_5, K_3, and K_3. Let G be the connected graph formed by adding two edges. Then G has Betti number 8, while $\gamma_M(G) = 3$. Although $X_1(G) = -3/2$, $\gamma(G) = 1$.

Another example is provided by replacing the K_5 in the last example by K_9. We have $X_1(G) = 5/6$, while $\gamma(G) = 3$. Since $\beta(G) = 30$ and $\gamma_M(G) = 14$, the graph is in set D.

By generalizing the examples in sets B and D in an obvious manner, we can construct infinite classes of graphs in each set.

4. <u>Upper imbeddable graphs and blocks</u>. We first show that the block which is the Petersen Graph is upper imbeddable.

REMARK 1. <u>The Petersen graph</u> P <u>is upper imbeddable</u>.

<u>Proof</u>. Label the graph P as below:

P:

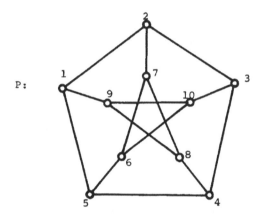

The following vertex permutations in an Edmonds' permutation scheme produce an imbedding of P with one face of length 30:

P_1: (5 2 9) P_6: (7 10 5)

P_2: (1 7 3) P_7: (2 8 6)

P_3: (10 4 2) P_8: (4 7 9)

P_4: (5 8 3) P_9: (8 1 10)

P_5: (1 4 6) P_{10}: (6 3 9)

Hence P is upper imbeddable.

One notices that many of the more well known blocks (e.g. K_n and $K_{m,n}$, for $m,n \geq 2$) are upper imbeddable. In [5] an example of a block which is not upper imbeddable is given. Here we give a method by which an infinite number of non-upper imbeddable blocks may be constructed.

THEOREM 3. Let G' be a graph with n components, G_1, G_2, ... G_n where n is at least three. For i ranging from 1 to (n-1), let e_i be an edge joining G_i to G_{i+1}. Let G = G' + Σe_i, i = 1, ..., n-1. Form H by adding an edge from G_1 to G_n. Then H is a non-upper imbeddable block if each of the G_i is a non-trivial block and any of the following four statements hold:

i) All of the G_i are cycles;

ii) At least two of the G_i are non-upper imbeddable;

iii) At least four of the G_i have odd Betti number;

iv) At least two G_i have odd Betti number while some other is non-upper imbeddable.

Proof. H is clearly a block. By calculation, we obtain $\beta(H) - 1 = \Sigma\beta(G_i)$. Furthermore, $\gamma_M(G) = \Sigma \gamma_M(G_i)$, (i = 1, ..., n), by the maximum genus summability theorem. Since adding an edge can increase maximum genus by at most one, we have $\gamma_M(H) \leq (\Sigma \gamma_M(G_i)) + 1$.

If all the G_i are cycles, then $\gamma_M(H) \leq 1$. Since $\beta(H) \geq 4$, H is not upper imbeddable.

Suppose G_1 and G_2 are not upper imbeddable. Then $\gamma_m(G_j) < \left\lceil \dfrac{\beta(G_j)}{2} \right\rceil$, j = 1, 2. Thus $\gamma_M(G_j) \leq \dfrac{\beta(G_j)}{2} - 1$, j = 1, 2. Hence,

$$\gamma_M(H) \leq \sum \frac{\beta(G_i)}{2} - 1 = \frac{\beta(H)}{2} - \frac{3}{2} < \left\lceil \frac{\beta(H)}{2} \right\rceil .$$

If G_1, G_2, G_3, and G_4, have odd Betti number, then $\gamma_M(G_j)$ is smaller than or equal to

$$\left\lceil \frac{\beta(G_j)-1}{2} \right\rceil ,$$

j = 1, 2, 3, 4. Hence, the inequalities in the last paragraph apply and H is not upper imbeddable.

If G_1 is not upper imbeddable, while G_2, G_3 have odd Betti number, then we obtain the same inequalities as before. Thus, H is not upper imbeddable under any of the stated conditions.

As mentioned earlier, when one examines blocks in relation to maximum genus, he soon discovers a certain prevalence of upper imbeddable graphs. We thus ask if all non-upper imbeddable blocks are as described in the theorem.

CONJECTURE. Let $\lambda(B)$ be the line connectivity of a block B. Then B is non-upper imbeddable if and only if $\lambda(B) = 2$.

We next discuss the relationship between upper imbeddable blocks of a graph and the upper imbeddability of the graph itself. A theorem in [5] can be rewritten as follows:

THEOREM. Let G be a connected graph all of whose blocks are upper imbeddable with even Betti number. Then G is upper imbeddable with even Betti number.

Several of the examples in Theorem 2 show the following remark true:

REMARK. A connected graph all of whose blocks are upper imbeddable need not be upper imbeddable.

When one examines the examples upon which the remark depends, he notices that each has at least one bridge. We therefore make the following conjecture:

CONJECTURE. If G is a connected bridgeless graph each of whose blocks is upper imbeddable, then G is upper imbeddable.

The concepts of upper and lower imbeddable graphs seem well worth further exploration. If such exploration ensues, this paper will have served its purpose.

References

1. J. Battle, F. Harary, Y. Kodama, and J. W. T. Youngs, Additivity of the genus of a graph, *Bull*. *Amer*. *Math*. *Soc*. 68 (1962), 565-568.

2. L. Beineke and F. Harary, Inequalities involving the genus of a graph and its thickness, *Proc*. *Glasgow* *Math*. *Assoc*. 7 (1965), 19-21.

3. R. Duke, How is a graph's Betti number related to its genus? *Amer*. *Math*. *Monthly* 78 (1971), 386-388.

4. R. Duke, The genus regional number and Betti number of a graph, *Canad*. *J*. *Math*. 18 (1966), 817-822.

5. E. Nordhaus, B. Stewart, and A. White, On the maximum genus of a graph, *J*. *Combinatorial* *Theory*, Ser. B, 11 (1971), 258-267.

6. E. Nordhaus, R. Ringeisen, B. Stewart, and A. White, A Kuratowski type theorem for the maximum genus of a graph, *J*. *Combinatorial* *Theory*, to appear.

7. R. Ringeisen, Determining all compact orientable 2-manifolds upon which $K_{m,n}$ has 2-cell imbeddings, *J*. *Combinatorial* *Theory*, Ser. B, 12 (1972), 101-104.

8. _____, Graphs of given genus and arbitrarily large maximum genus, submitted for publication.

9. G. Ringel, Das Geschlecht des vollständigen paaren graphen, *Abh*. *Math*. *Sem*. *Univ*. *Hamburg* 28 (1965), 139-350.

10. _____ and J. W. T. Youngs, Solution of the Heawood map-coloring problem, *Nat*. *Acad*. *Sci*. 60 (1968), 438-455.

TRIANGULAR EMBEDDINGS OF GRAPHS*

Gerhard Ringel
University of California
Santa Cruz, CA 95060

1. _Introduction._ The proof of Heawood's formula for the chromatic number of a 2-dimensional compact manifold (closed surface) different from the sphere is founded on the determination of the genus and the non-orientable genus of the complete graph K_n [3] or [5].

In order to do this one has to construct embeddings of K_n or other graphs G which are similar to K_n into a suitable closed surface S such that each face is a triangle. We call this a triangular embedding and use the notation $G \lhd S$.

Gustin [1] invented a powerful method to construct triangular embeddings. He used current graphs satisfying Kirchhoff's Current Law. His method was completed by some additional ideas by J. W. T. Youngs and the author. Youngs used the very good idea of "cascades" in order to construct non-orientable triangular embeddings of graphs. But even the cascades we can now eliminate by better methods in some cases. All these ideas we will explain in this lecture by means of certain examples. We also present a new and elegant non-orientable triangular embedding of K_n for $n \equiv 3$ or $7 \pmod{12}$.

2. _Notation and results._ Denote the orientable (non-orientable) closed surface of genus $p(q)$ by $S_p(N_q)$. If S is a closed surface, denote the Euler characteristic of S by $E(S)$. Let G be a graph with α_0 vertices and α_1 arcs.

The following theorems are well known. For more details see J. W. T. Youngs [7].

THEOREM 1. _If_ G _is embeddable into_ S, _then_ $\alpha_1 \leq 3\alpha_0 - 3E(S)$.

THEOREM 2. _If_ $G \lhd S$, _then_ $\alpha_1 = 3\alpha_0 - 3E(S)$.

---

* Research supported by the National Science Foundation.

If a graph G is embeddable into S_p but not into S_{p-1}, then p is called the genus of G. We write $p = \gamma(G)$.

THEOREM 3. If $G \lhd S_p$, then $\gamma(G) = p$.

The proof of the formula

$$(1) \qquad \gamma(K_n) = \left\{ \frac{(n-3)(n-4)}{12} \right\}, \quad \text{for} \quad n \geq 3,$$

is based in a certain natural way on the existence of the following triangular embeddings.

a). For $n \equiv 0, 3, 4$ or 7 (mod 12) there exists a triangular embedding of K_n into a suitable orientable surface.

For all other residue classes n (mod 12) a triangular orientable embedding of K_n does not exist, because of Theorem 2 and the fact that the Euler characteristic of an orientable surface is even.

If we remove one arc out of the complete graph K_n, we denote the resulting graph by $K_n - K_2$. Similarly $K_n - K_3$ is the graph obtained from K_n by removing 3 arcs forming a triangle.

b). For $n \equiv 1, 6, 9, 10$ (mod 12) and $n \geq 9$, $n \neq 13$, there exists a triangular embedding of the graph $K_n - K_3$ into an orientable surface.

The case $n = 6$ is a remarkable exception. One can show that $K_6 - K_3$ is not triangularly embeddable into the sphere. For $n = 13$ it is unknown whether a triangular embedding of $K_{13} - K_3$ exists. But (1) can be proved for $n = 13$ by another method.

c). For $n \equiv 2$ or 5 (mod 12) but $n \not\equiv 2$ (mod 24), there exists a triangular embedding of $K_n - K_2$ into an orientable surface.

Unfortunately for all $n \equiv 2$ (mod 24) a proof of c) is missing, although again (1) can be proved by another method.

d). A triangular embedding of $K_n - K_5$ exists for $n \equiv 11$ (mod 12) with $n \geq 23$ and it probably exists for $n \equiv 8$ (mod 12) with $n \geq 20$.

It can be shown that $K_8 - K_5$ is not triangularly embeddable.

For non-orientable surfaces the picture is much easier.

α). For $n \equiv 0$ or 1 (mod 3) and $n \neq 7$, $n \geq 6$, there exists a triangular embedding of K_n into a non-orientable surface.

β). <u>For</u> n ≡ 2 (mod 3) <u>and</u> n ≥ 5, n ≠ 8, <u>there exists a tri-angular embedding of</u> $K_n - K_2$ <u>into a non-orientable surface</u>.

There are two remarkable exceptions:

It is proved that triangular non-orientable embeddings for K_7 and $K_8 - K_2$ do not exist. (See Ringel [4].)

EXAMPLES. For Figure 1 opposite points of the circle are to be identified. Then an embedding $K_6 \triangleleft N_1$ is obtained. Figures 2 and 3 illustrate embeddings $K_5 - K_2 \triangleleft S_0$ and $K_7 \triangleleft S_1$.

3. <u>Rotations of graphs</u>. Given a graph G, a rotation at a vertex i of G is an oriented cyclic permutation of all the arcs incident with i. If, for instance, the valence of i is three, then there are two different rotations on i. A rotation on G is a rotation on each vertex of G.

If a graph is represented on the plane and has vertices of valences ≤ 3, then each rotation can be given as a clockwise or counterclockwise reading of the arcs incident with the considered vertex. In the figures, if the reading is to be clockwise (counterclockwise) the vertex is represented by a small filled-in (empty) circle.

We consider now certain closed walks in a graph which are determined by a given rotation of the graph. In order to explain this easily, consider the graph as a road-map, the arcs as (two-way) highways, and each vertex as a one-way traffic circle whose rotation determines the direction of the flow of traffic. Assume a traveller takes a trip starting at vertex i and continuing on highway ij, where each time he reaches the next vertex, regarded as a traffic circle, he takes the first exit. (This of course, puts him on another highway if the valence of the vertex is ≥ 2.)

We call this trip the circuit generated by the vertex i and the arc ij. In this circuit it may happen that vertices or arcs are repeated, although each time an arc is repeated, the repetition occurs in the opposite direction. So the rotation of the graph, as given in Figure 4, induces one single circuit (Figure 5). In general a rotation induces a certain number of circuits. If the graph is a tree each rotation induces only one single circuit.

A rotation of a graph can be presented by a combinatorial scheme in the following way. Denote each vertex by an integer; then write for each vertex i the adjacent vertices in exactly the oriented cyclic order which is given by the rotation. For instance the scheme

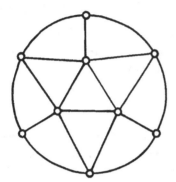

Figure 1.

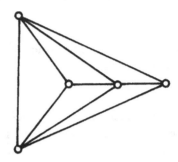

Figure 2.

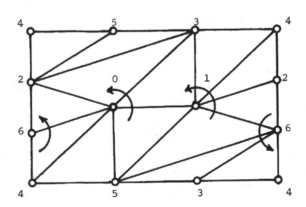

Figure 3.

$$
\begin{array}{llllllll}
0. & 1 & 3 & 2 & 6 & 4 & 5 \\
1. & 2 & 4 & 3 & 0 & 5 & 6 \\
2. & 3 & 5 & 4 & 1 & 6 & 0 \\
3. & 4 & 6 & 5 & 2 & 0 & 1 \\
4. & 5 & 0 & 6 & 3 & 1 & 2 \\
5. & 6 & 1 & 0 & 4 & 2 & 3 \\
6. & 0 & 2 & 1 & 5 & 3 & 4
\end{array}
\tag{2}
$$

represents a rotation of K_7. This particular rotation we have obtained from the embedding $K_7 \lhd S_1$ illustrated in Figure 3, in the obvious way.

One can easily check that the scheme (2) satisfies the

RULE Δ^*. If in line i one has i. ... jk. ..., then in line k one must have k. ... ij. ... .

The rule Δ^* really reflects the fact that the embedding $K_7 \lhd S_1$ is triangular. On the other hand it is known (Theorem of Edmonds [2]) that, if the scheme of a rotation of a graph G satisfies rule Δ^*, then there exists a triangular embedding of G into a suitable orientable surface; in particular rule Δ^* says that each induced circuit has length 3.

Consider the used symbols 0, 1, 2, 3, 4, 5, 6 as the elements of the cyclic group Z_7. We observe that in the scheme (2) the row i can be obtained by adding i to each element in row 0 with no change in order. We say that in the scheme (2) the additive rule holds. It really means that the automorphism group of the polyhedron given by $K_7 \lhd S_1$ has a cyclic subgroup of order 7.

4. Current graphs. Now we are going to explain the method of current graphs. We wish to construct a triangular embedding $K_{19} \lhd S_{20}$. Denote the vertices of K_{19} by the elements of the group Z_{19}. The problem is to find a permutation of the elements 1, 2, ..., 18, that we can use as row 0. It should be such that, if we construct the other rows just by the additive rule, we obtain a scheme satisfying rule Δ^*.

Let us write the nonzero elements of the group Z_{19} in the form $\pm 1, \pm 2, ..., \pm 9$. Then consider the graph of Figure 4 and assign to each arc one of the group elements 1, 2, ..., 9 (it is the current of the arc), and choose for each arc a certain direction indicated by an

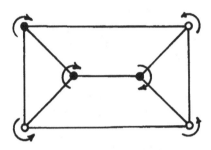

Figure 4.

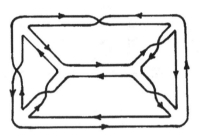

Figure 5.

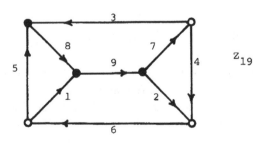

Z_{19}

Figure 6.

arrow. The resulting figure we call a _current graph_. The current
graph of Figure 6 has the properties (we will call them construction
principles):

C1). If a_1, a_2, ..., a_r are all the currents, where r is the
number of arcs, then 0, $\pm a_1$, $\pm a_2$, ..., $\pm a_r$ are all elements of the
group.

C2). At each vertex of valence 3 the sum of the inward flowing
currents equals the sum of the outward flowing currents (Kirchhoff's
Current Law).

C3). There is a rotation of the current graph (considered just
as a graph) which induces exactly one circuit.

We consider now the circuit (Figure 5) induced by the rotation
given in Figure 4. We propose to keep a _log_ of this circuit. The log
will consist of the succession of the currents of the arcs along which
the traveller moves. If he moves on the arc with current a but in
the opposite direction then we write in the log the element -a. This
way we get the succession:

$$7, 4, -2, -9, -1, 5, -3, -7, 2, 6, 1, -8, -5, -6, -4, 3, 8, 9.$$

We take this as the row 0, omit the commas, and write each element
of Z_{19} in the positive form:

$$0. \qquad 7 \; 4 \; 17 \; 10 \; 18 \; 5 \; 16 \; 12 \; 2 \; 6 \; 1 \; 11 \; 14 \; 13 \; 15 \; 3 \; 8 \; 9$$

By the additive rule we obtain from row 0 the row i (i = 1,
2, ..., 18). It is very easy to prove that the scheme we obtained
satisfies rule Δ^*. It is an immediate consequence of Kirchhoff's
Current Law (C3). This scheme guarantees the existence of a triangular embedding of K_{19} into an orientable surface. It is S_{20},
which is easy to prove using Euler's formula $\alpha_0 - \alpha_1 + \alpha_2 = 2 - 2_p$.
There is an easy generalization to all graphs K_n with $n \equiv 7$
(mod 12). (See Youngs [8].)

For all other cases one has to extend the list of construction
principles. We explain it in another example. Consider the group Z_6
and the current graph in Figure 7. Observe that the properties C1),
C2), C3) are valid. The arc (one can consider it as a loop) with the
number 3 has no direction because 3 = -3 in this group.

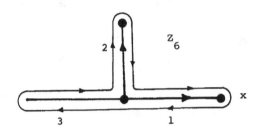

Figure 7.

The log of this current graph reads

$$2 \quad 4 \quad 1 \quad x \quad 5 \quad 3 \quad .$$

The symbol x is not a group element. But we record it in the log
when it appears on a vertex. For those extra symbols x we define
the addition $a + x = x$ for each element a of the group. Take the
log of the circuit as the row 0 and produce the other rows by additive rule. The scheme

$$
\begin{array}{lllllll}
0. & 2 & 4 & 1 & x & 5 & 3 \\
1. & 3 & 5 & 2 & x & 0 & 4 \\
2. & 4 & 0 & 3 & x & 1 & 5 \\
3. & 5 & 1 & 4 & x & 2 & 0 \\
4. & 0 & 2 & 5 & x & 3 & 1 \\
5. & 1 & 3 & 0 & x & 4 & 2 \\
\end{array}
$$

which is obtained is not yet complete. But the row x can be manu-
factured by using rule Δ^*:

$$x. \quad 0 \quad 1 \quad 2 \quad 3 \quad 4 \quad 5 \quad .$$

The total scheme of all 7 rows satisfies rule Δ^* and therefore
it represents an embedding $K_7 \lhd S_1$. In fact, it is the same embedding

as the one given by the scheme (2). This time we have seen that the automorphism group of the polyhedron $K_7 \triangleleft S_1$ contains a cyclic subgroup of order 6 also.

Observe that the current graph of Figure 7 obeys the following additional construction principles. We call an arc incident with a vertex of valence one an _end arc_.

C4). _An element of order_ 2 _in the group must be the current of an end arc, which may be considered as a loop._

C5). _An element of order_ 3 _can be the current of an end arc._

C6). _A generator of the group can be the current of an end arc, but then an extra symbol such as_ x _has to be on the vertex of valence one incident with the end arc._

It is easy to prove that, if the current graph satisfies the six properties C1) to C6), then the scheme constructed in the described way satisfies rule Δ^*. Especially in the case of C6) one can complete the scheme by a row x.

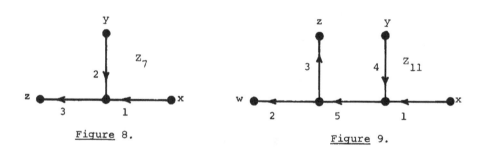

Figure 8.   Figure 9.

In the two examples of Figures 8 and 9, all properties C1) to C6) hold. These current graphs lead directly to triangular orientable embeddings for $K_{10} - K_3$ and for $K_{15} - K_4$, respectively.

Unfortunately, it is not possible to use the same method to find triangular orientable embeddings for $K_{13} - K_3$ and $K_{20} - K_5$ because in these two cases there are not enough generators in the group.

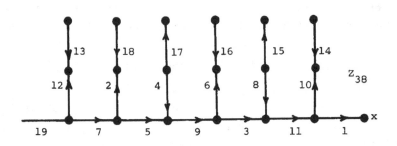

Figure 12.

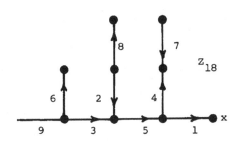

Figure 13.

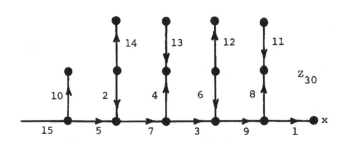

Figure 14.

5. **Remarks on non-orientable embeddings.** When we translate the whole method to the case of non-orientable surfaces, we have to replace the rule Δ^* by the weaker

RULE Δ: **If in the row** i **one sees** i. ... jk. ..., **then in the row** k **one has** k. ... ij. ... **or** k. ... ji. ... **and in the row** j one has j. ... ik. ... **or** j. ... ki. ... .

It is easy to observe that rule Δ is equivalent to the

RULE R: **If in the row** i **one sees** i. ... jkℓ ..., **then in the row** k **one has** k. ... ℓij ... **or** k. ... jiℓ. ... . **For instance the scheme**

$$
\begin{array}{llllll}
0. & 1 & 4 & 3 & x & 2 \\
1. & 2 & 0 & 4 & x & 3 \\
2. & 3 & 1 & 0 & x & 4 \\
3. & 4 & 2 & 1 & x & 0 \\
4. & 0 & 3 & 2 & x & 1 \\
x. & 4 & 2 & 0 & 3 & 1
\end{array}
\tag{3}
$$

satisfies both rules Δ and R. It is shown (Ringel [3]) that a scheme satisfying rule R represents a triangular embedding into a closed surface (which may or may not be orientable). In each case the non-orientability must be shown by an extra checking. For instance the scheme (3) represents an embedding $K_6 \triangleleft N_1$.

We have obtained the scheme (3) from the current graph of Figure 10 in the described way. The current graph of Figure 10 (and of all

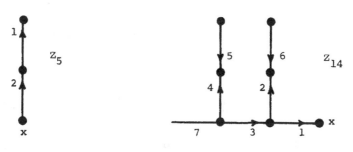

Figure 10.   Figure 11.

the later figures) has the pertinent property (called doubling):

C7). *If a vertex i is of valence two then one arc incident with i must be an end arc. If b is the current of this end arc, then 2b must be the current of the other arc incident with i pointing in the same direction (or -2b pointing in the opposite direction).*

The vertex of valence one on the end of this end arc is not allowed to carry an extra letter, like x.

One can show that a current graph satisfying all 7 construction principles C1) to C7) leads to a scheme satisfying Rule R. Moreover, if C7) is used at least once, then the obtained triangular embedding is non-orientable.

One can check that the current graphs in Figures 11, 12, 13, and 14 satisfy all seven construction principles C1) to C7). Therefore Figures 11 and 12 represent a non-orientable triangular embedding of K_{15} and of K_{39}, respectively. The general pattern is obvious (easy arithmetic sequences) and gives a triangular non-orientable embedding of K_{12s+3}.

The Figures 13 and 14 illustrate a triangular non-orientable embedding of K_{12s+7} for $s = 1$ and 2. Again the generalization for arbitrary s is easy to see. These two solutions in the non-orientable cases 3 and 7 (mod 12) are very new and are not published elsewhere.

References

1. W. Gustin, Orientable embedding of Cayley graphs, *Bull. Amer. Math. Soc.* 69 (1963), 272-275.

2. J. Edmonds, A combinatorial representation for polyhedral surfaces, *Notices Amer. Math. Soc.* 7 (1960), 646.

3. G. Ringel, *Färbungsprobleme auf Flächen und Graphen*, VEB Deutscher der Wissenschaften, Berlin, 1959.

4. G. Ringel, Wie man die geschlossenen nichtorientierbaren Flächen in möglichst wenig Dreiecke zerlegen kann, *Math. Ann.* 130 (1955), 317-326.

5. G. Ringel and J. W. T. Youngs, Solution of the Heawood map-coloring problem, *Proc. Nat. Acad. Sci. USA* 60 (1968), 438-445.

6. _____, Solution of the Heawood map-coloring problem, Case 11, *J. Combinatorial Theory* 7 (1969), 71-93.

7. J. W. T. Youngs, Minimal imbeddings and the genus of a graph, _J. Math. Mech._ 12 (1963), 303-315.

8. _____, Solution of the map-coloring problem -- Cases 1, 7, and 10. _J. Combinatorial Theory_ 9 (1970), 220-231.

PROGRESS ON THE PROBLEM OF ECCENTRIC HOSTS

Seymour Schuster*
Carleton College
Northfield, MN 55057

1. **Introduction.** Imagine an eccentric married couple who wish
to issue invitations to a dinner party at which there will be, among
the many dining tables, one special circular table that seats m people and another that seats n people. The wife would be very pleased
is she could be sure to find m people at the party who could be
seated at the first special table so that each one would require no
introduction to the persons seated on his immediate right and left.
The husband, on the other hand, would be delighted if he could find n
people who could be seated at the second special table so that no one
at the table knew either of his neighbors. However, since the hosts
are happily married, they would gladly settle for a party that satisfied either the wife's or the husband's eccentric taste. Thus, they
seek the smallest number p of people to be present at the party so
that either there are m people who could satisfy her condition or
there are n people present who could satisfy his condition.

This problem of the eccentric hosts admits a graph theoretic interpretation that has received much attention during the past year.
For $m,n \geq 3$, let $c(m,n)$ be the smallest integer p such that for
any graph G of order p, either G contains an m-cycle or its complement $\bar{G}$ contains an n-cycle. The numbers $c(m,n)$ are referred to
as cycle Ramsey numbers since they are obvious analogues of the celebrated Ramsey numbers $r(m,n)$; the definition of $c(m,n)$ calls for
the existence of cycles in complementary graphs, whereas $r(m,n)$ calls
for the existence of complete subgraphs in the complementary graphs.
The value of $c(m,n)$ is the solution to the problem of our eccentric
hosts.

What appears to be the sharpest conjecture (see [7]) for the
values of the cycle Ramsey numbers is the following.

---

*Research partially supported by a grant from the Alfred P. Sloan
Foundation to Carleton College.

CONJECTURE. (a) <u>If</u> $3 \leq m \leq n$ <u>and</u> m <u>is</u> <u>odd</u>, <u>then</u>
$$c(m,n) = 2n - 1, \quad \underline{except} \ \underline{for} \quad (m,n) = (3,3).$$

(b) <u>If</u> $3 < m \leq n$ <u>and</u> m <u>is</u> <u>even</u>, <u>then</u>

$$c(m,n) = \begin{cases} n + \frac{m}{2} - 1 & \underline{if} \quad n \geq \frac{3}{2} m \\ n + \frac{m}{2} - 1 & \underline{if} \quad n \ \underline{is} \ \underline{even} \ and \ n \leq \frac{3}{2} \\ 2m - 1 & \underline{if} \quad n \ \underline{is} \ \underline{odd} \ \underline{and} \ n \leq \frac{3}{2} m, \end{cases}$$

<u>except</u> <u>for</u> $(m,n) = (4,4)$.

It is the purpose of this paper to summarize the progress to date towards proving the conjecture and to present three new cycle Ramsey numbers, namely $c(6,7)$, $c(6,8)$, and $c(6,9)$. Only outlines of the lengthy proofs will be given here, with the anticipation that there will be an opportunity to publish complete proofs elsewhere at some subsequent date.

2. <u>Summary</u> <u>of</u> <u>previous</u> <u>results</u>. In the following statements we shall always take $3 \leq m \leq n$. This stipulation nets an economy in our tabulation with no loss of information (due to the fact that $c(m,n) = c(n,m)$). The reference(s) to the right of each result indicates where the result can be found.

Precise <u>values</u> <u>of</u> $c(m,n)$.

(1) $c(3,3) = 6$ (Very old See, e.g., [4, p. 16].)

(2) $c(3,4) = 7$ [3], [5], and [6]

(3) $c(3,n) = 2n - 1$, if $n \geq 4$ [5] and [6]

(4) $c(4,4) = 6$ [2], [5], and [6]

(5) $c(4,5) = 7$ [5] and [6]

(6) $c(4,n) = n + 1$, if $n \geq 6$ [5] and [6]

(7) $c(5,n) = 2n - 1$, if $n \geq 5$ [5] and [6]

(8) $c(6,6) = 8$ [6]

(9) If m is odd, $c(m,n) = 2n - 1$
for $n > \frac{1}{2}(m^2 + m)$.   [1]

(10) If m is odd, $c(m,m) = 2m - 1$.   [1] and [7]

(11) If m is even, $c(m,n) = n + \frac{m}{2} - 1$
for $n \geq m^2 - \frac{m}{2} + 2$.   [1]

Upper bounds for $c(m,n)$.

(12) $c(m,n) \le 2n - 1$ for $n > \frac{1}{2}(m^2 + m)$ [1]

(13) $c(m,m) \le 2m - 1$ [1]

Lower bounds for $c(m,n)$.

(14) If m is odd, $c(m,n) \ge 2n - 1$.   [1] and [7]

(15) If m is even

$$c(m,n) \ge \begin{cases} n + \frac{m}{2} - 1 \text{ if } n \ge \frac{3}{2}m \\ n + \frac{m}{2} - 1 \text{ if } n \text{ is even} \\ \qquad \text{and } n \le \frac{3}{2}m \\ 2m - 1 \text{ if } n \text{ is odd} \\ \qquad \text{and } n \le \frac{3}{2}m. \end{cases}$$
     [7]

3. **New results.** We begin by giving the cycle Ramsey number $c(6,7)$.

THEOREM. The cycle Ramsey number $c(6,7) = 11$.

Outline of proof. It follows from (15) that $c(6,7) \ge 11$. Therefore, we let G be an arbitrary graph of order 11, and assume that G contains no 6-cycle. Since $c(6,6) = 8$ (see (8) above), we know that $\overline{G}$ contains a 6-cycle $C: u_1, u_2, u_3, u_4, u_5, u_6, u_1$. Denote the remaining vertices of $\overline{G}$ (and hence G) by v_1, v_2, v_3, v_4, and v_5. If any v_i is adjacent in $\overline{G}$ to two consecutive vertices of C, then $\overline{G}$ contains a 7-cycle, completing the proof. Suppose, then, that no such v_i exists.

Each v_i is adjacent in G to at least one vertex of the pair $\{u_1, u_2\}$; hence, at least three of the five v_i are adjacent in G to the same one of $\{u_1, u_2\}$. Without loss of generality, we stipulate that $v_1, v_2,$ and v_3 are adjacent in G to u_1.

Further, two vertices of $\{v_1, v_2, v_3\}$ are adjacent in G to at least one of the vertices of $\{u_2, u_3\}$. Again without loss of generality, we specify that it is v_1 and v_2 that are adjacent to a common u_j, $j = 2$ or $j = 3$. Thus, we are led to consider two cases resulting from $j = 2$ and $j = 3$.

Case 1. Suppose that v_1 and v_2 are adjacent in G to u_2. Keeping in mind that no v_i is adjacent in $\overline{G}$ to consecutive vertices of C, we can use the pigeonhole principle repeatedly to determine which edges $v_i u_j$ are in $E(G)$ and which are in $E(\overline{G})$. The analysis shows that G contains a 6-cycle, in contradiction to our first assumption concerning G.

Case 2. Suppose v_1 and v_2 are adjacent in G to u_3. Each of the vertices v_3, v_4, and v_5 is adjacent in G to at least two vertices of $\{u_4, u_4, u_5, u_6\}$. Hence, at least two of the v_i, $3 \leq i \leq 5$, are adjacent to the same u_j, $3 \leq j \leq 6$. This forces us into eight subcases: (a) - (d), in which v_3 and v_4 are adjacent in G to u_j, $j = 3$, 4, 5, and 6; and (e) - (h), in which v_4 and v_5 are adjacent in G to u_j, $j = 3$, 4, 5, and 6. (Coupling v_3 and v_5 would be superfluous because a mere relabeling would reduce such cases to ones already considered.)

Each of the subcases (a) - (h) also leads to the existence of a contradictory 6-cycle in G. Hence, $\overline{G}$ must possess a 7-cycle.

THEOREM. The cycle Ramsey number $c(6,8) = 10$.

Outline of proof. Since (15) states that $c(6,8) \geq 10$, we consider the graph G to be of order 10 and without a 6-cycle. The fact that $c(6,6) = 8$ assures us that $\overline{G}$ contains a 6-cycle: $C = u_1$, u_2, u_3, u_4, u_5, u_6, u_1. The remaining vertices are designated v_1, v_2, v_3, and v_4.

The proof is divided into two major parts, each of which requires a further breakdown into cases. Part I assumes that neither G nor $\overline{G}$ contains a 7-cycle. Part II will assume that either G or $\overline{G}$ does possess a 7-cycle.

Part I. Since $\overline{G}$ has no 7-cycle, no vertex v_i, $1 \leq i \leq 4$, is adjacent in $\overline{G}$ to two consecutive vertices of C. Hence, each vertex v_i is adjacent in G to at least three of the vertices u_j, $i \leq j \leq 6$. We are therefore forced to consider two cases.

Case A. Suppose at least three of the vertices v_i are adjacent in G to a single u_j. For definiteness, we take $v_1 u_1$, $v_2 u_1$, $v_3 u_1 \in E(G)$. Since no v_i may be adjacent in $\overline{G}$ to two consecutive vertices of C, we know that two of the vertices of $\{v_1, v_2, v_3\}$ are adjacent in G to u_2 or u_3. Thus we consider two subcases within Case A.

Subcase Al. Assume v_1 and v_2 are adjacent in G to u_2. We note that v_1 is adjacent in G to at least one of the vertices of $\{u_4, u_5\}$. In either case ($v_1u_4 \in E(G)$ or $v_1u_5 \in E(G)$), an 8-cycle appears in $\overline{G}$.

Subcase A2. Assume v_1 and v_2 are adjacent in G to u_3. As in Subcase Al, v_1 is adjacent in G to at least one of the vertices $\{u_4, u_5\}$. If $v_1u_4 \in E(G)$, we obtain the existence of a contradictory 6-cycle in G; if $v_1u_5 \in E(G)$, we find that $\overline{G}$ contains an 8-cycle.

Case B. Suppose at most two of the vertices v_i, $1 \le i \le 4$, are adjacent in G to a single vertex u_j, $1 \le j \le 6$. For definiteness we take v_1u_1, $v_2u_1 \in E(G)$ and v_3u_1, $v_4u_1 \in E(\overline{G})$. This case, also, leads to an 8-cycle in $\overline{G}$.

Part II. If it is G that contains a 7-cycle w_1, w_2, w_3, w_4, w_5, w_6, w_7, w_1, then alternate vertices of this cycle must be joined in $\overline{G}$. Hence, $\overline{G}$ also contains the 7-cycle, namely w_1, w_3, w_5, w_7, w_2, w_4, w_6, w_1. So we may safely assume that $\overline{G}$ possesses a 7-cycle, which we designate C: u_1, u_2, ..., u_7, u_1; we call the remaining vertices v_1, v_2, and v_3.

Since no vertex v_i is adjacent in $\overline{G}$ to two consecutive vertices of C, we have that each vertex v_i is adjacent in G to at least four vertices u_j, $1 \le j \le 7$. Thus, every two vertices v_i, $1 \le i \le 3$, are joined in G to at least one common vertex u_j.

It cannot happen that every two vertices v_i are joined to only one common vertex u_j. This follows from the pigeonhole principle. Further, it cannot happen that two vertices v_i are mutually adjacent in G to three vertices u_j; for, again by the pigeonhole principle, we see that a 6-cycle would occur in G. Thus, we need consider only cases in which two vertices v_i are mutually adjacent to two vertices u_j.

Case A. Suppose v_1 and v_2 are adjacent in G to two consecutive vertices of C, say u_1 and u_2, and that v_1 and v_2 are not mutually adjacent to any other vertices u_j, $3 \le j \le 7$. One of the vertices of $\{v_1, v_2\}$ is adjacent in G to a vertex of C consecutive with u_1 or u_2, for otherwise v_1 and v_2 would both be adjacent in G to three distinct vertices u_j. Without loss of generality, we take $v_1u_3 \in E(G)$. This allows us to determine whether the various edges $v_i u_j$, $i = 1, 2$ and $j = 3, 4, 5, 6, 7$, are in $E(G)$ or $E(\overline{G})$.

We observe that vertex v_3 is adjacent in G to one of the vertices of $\{u_1, u_2\}$ and also to some other vertex u_j, $3 \leq j \leq 7$. This information is sufficient to conclude that in this case a 6-cycle exists in G.

Case B. Suppose v_1 and v_2 are both adjacent in G to two u_j which are two units apart on C, and that v_1 and v_2 are not mutually adjacent to any other u_j. Arguing as we did in Case A, we arrive at the same result, namely a 6-cycle in G.

Case C. Suppose the vertices v_1 and v_2 are mutually adjacent in G to two vertices u_j which are three units apart on C, and that v_1 and v_2 are not mutually adjacent to any other vertex u_j. This case also results in the existence of a contradictory 6-cycle in G.

Hence, all cases lead to an 8-cycle in $\overline{G}$ or a contradiction, so we conclude that $c(6,8) = 10$.

THEOREM. The cycle Ramsey number $c(6,9) = 11$.

Outline of proof. From (15) we know that $c(6,9) \geq 11$. Therefore, we let G be an arbitrary graph of order 11, and we assume that G has no 6-cycle. Since $c(6,8) = 10$ by the previous theorem, we know that G contains an 8-cycle $C: u_1, u_2, \ldots, u_8, u_1$. The remaining vertices shall be denoted $v_1, v_2,$ and v_3.

If the vertex v_i is adjacent in $\overline{G}$ to two consecutive vertices of C, a 9-cycle exists in $\overline{G}$ and the proof is completed. Hence, we need consider only cases in which no vertex v_i is adjacent in $\overline{G}$ to two consecutive vertices of C. It follows that each vertex v_i is joined in G to at least four vertices u_j, $1 \leq j \leq 8$. We consider two cases.

Case A. Suppose that two of the vertices v_i, say v_1 and v_2, are not simultaneously adjacent in G to any of the vertices u_j. For definiteness, we take $v_1 u_{2k-1} \in E(G)$ and $v_2 u_{2k} \in E(G)$, for $k = 1, 2, 3,$ and 4. The vertex v_3 is adjacent in G to at least two of the vertices of $\{u_1, u_3, u_5, u_7\}$ or at least two of the vertices of $\{u_2, u_4, u_6, u_8\}$. Without loss of generality, we assume that v_3 is adjacent in G to at least two of the vertices of

$\{u_1, u_3, u_5, u_7\}$ and then consider subcases according as there are four, three, or two such adjacencies.

Subcase A1. Suppose $v_3u_{2k-1} \in E(G)$, for $k = 1, 2, 3$, and 4. Under this assumption, we find that $\overline{G}$ contains a 9-cycle.

Subcase A2. Suppose the vertex v_3 is adjacent in G to three vertices of $\{u_1, u_3, u_5, u_7\}$. In this subcase, also, a 9-cycle results in $\overline{G}$.

Subcase A3. Suppose the vertex v_3 is adjacent to exactly two vertices of $\{u_1, u_3, u_5, u_7\}$. This subcase reduces to one of the previous two under the relabeling $v_1 \leftrightarrow v_2$, $u_{2k} \leftrightarrow u_{2k-1}$.

Case B. Every two of the vertices v_i, $i = 1, 2, 3$, are mutually adjacent in G to some vertex u_j, $1 \le j \le 8$. That is, there exist vertices u', u'', u''' of C such that v_1u', v_2u', v_2u'', v_3u'', v_1u''', $v_3u''' \in E(G)$. If there exist such u', u'', u''' that are distinct, then there clearly exists a 6-cycle in G. So, we henceforth assume coincidence among the available choices for u', u'', and u'''.

If there is a vertex, say u_1, which is simultaneously adjacent in G to the vertices v_1, v_2, and v_3, then there is another vertex $u_{1'}$ ($\ne u_1$) which is simultaneously adjacent in G to at least two of the vertices $\{v_1, v_2, v_3\}$. We consider two subcases according as $u_{1'}$ is adjacent in G to three or two of the vertices in the set $\{v_1, v_2, v_3\}$.

Subcase B1. Suppose $u_jv_i \in E(G)$ for $j = 1, 1'$ and $i = 1, 2$, and 3. Under this supposition, a 6-cycle exists in G.

Subcase B2. Suppose $u_1v_i \in E(G)$ for $i = 1, 2, 3$, and $u_{1'}v_1$, $u_{1'}v_2 \in E(G)$. Although it is immediate that $v_3u_{1'} \in E(\overline{G})$ (in order to avoid being in the previous subcase), it is not clear what position $u_{1'}$ holds in C. An analysis of the various possibilities is required, but in each case either $\overline{G}$ possesses a 9-cycle or G contains a 6-cycle. This concludes the proof that $c(6,9) = 11$.

While this author realized that the knowledge of three more cycle Ramsey numbers would make only a small contribution to the overall problem, he pursued these numbers through exhausting proofs in hopes of obtaining insight that would carry over to the more general cases—perhaps enough to determine $c(m,n)$ for m even. Regrettably, he

has come away with weakened vision -- but not with the depth of understanding that was sought. It is hoped that some graph-theoretician
colleague who has read this far can glean something from the above
proofs that will help him settle the conjecture.

Added in proof: A communication from P. Erdös to the Conference
reported that V. Rosta has recently shown that for m odd, $c(m,n) \leq$
$2n - 1$. This, together with (14) completes the proof that $c(m,n) =$
$2n - 1$ if m is odd. Also, since submitting this paper, I have
obtained a proof that $c(6,n) = n + 2$, for all $n \geq 8$.

References

1. J. A. Bondy and P. Erdös, Ramsey numbers for cycles in graphs,
    submitted for publication.

2. V. Chvátal and F. Harary, Generalized Ramsey theory for graphs II.
    Small diagonal numbers, Proc. Amer. Math. Soc., to appear.

3. V. Chvátal and F. Harary, Generalized Ramsey theory for graphs III.
    Small off-diagonal numbers, Pacific J. Math., to appear.

4. F. Harary, Graph Theory, Addison-Wesley, Reading (1969).

5. G. Chartrand and S. Schuster, On the existence of specified cycles
    in complementary graphs, Bull. Amer. Math. Soc. 77 (1971), 995-8.

6. G. Chartrand and S. Schuster, On a variation of the Ramsey number,
    submitted for publication.

7. G. Chartrand and S. Schuster, A note on cycle Ramsey numbers,
    submitted for publication.

MAGIC SETS

B. M. Stewart
Michigan State University
East Lansing, MI 48823

A way to study one system is to consider its relation to another
system. In my paper "Magic Graphs" [1], I show a way to relate a
finite graph G to a nest of vector spaces over a field F. (See
properties (M.1), (M.2) to follow.) A classification of graphs results from a study of the dimensions of these vector spaces. A
further classification results from a study of numerical properties of
the vectors in the vector spaces. (See properties (M.3), (M.4), (M.5)
to follow.) In this preliminary report I indicate how the same ideas
can be generalized to cover a wide variety of situations. From recent literature I select four problems of graph theory to which this
theory may be profitably applied. I conclude with an example from the
magic squares of number theory, a "diabolic doughnut", an old example
re-examined with the use of vector spaces.

If we are given a finite set $T = \{t_j\}$ and a field F, then the
set $A_F(T)$ of all functions α having the domain T and values in
F is a vector space, under the rule $(x_1\alpha_1 + x_2\alpha_2)(t) = x_1\alpha_1(t) +
x_2\alpha_2(t)$ for all t in T and all x_1, x_2 in F. It is readily
seen that $\dim A_F(T) = |T|$, since the functions $\{\varepsilon_i\}$ defined by
$\varepsilon_i(t_j) = \delta_{ij}$ form a linearly independent basis. The zero vector is
the function ζ with $\zeta(t) = 0$ for every t in T.

We suppose we are given a set Q of distinguished non-empty subsets Q_{ij} of T arranged in k classes as follows:

$$Q = \{Q_{11}, \ldots, Q_{1m_1}; Q_{21}, \ldots, Q_{2m_2}; \ldots; Q_{k1}, \ldots, Q_{km_k}\}.$$

To help fix ideas let us describe an example with $k = 2$. Consider a graph G with a polyhedral imbedding on some surface. Let T
be the vertices and edges of G. Let Q_{1j} be the set of edges inci-
dent to the vertex v_j and $m_1 = |V|$. Let Q_{2j} be the set of edges
and vertices incident to the face f_j and let m_2 be the number of
faces in this particular imbedding.

For each α in $A_F(T)$ and each Q_{ij} in Q define the subset sum

$$\sigma^\alpha(Q_{ij}) = \sum \alpha(t),$$

summed over all t in Q_{ij}. Note that

$$\sigma^{x_1\alpha_1 + x_2\alpha_2}(Q_{ij}) = \sum (x_1\alpha_1 + x_2\alpha_2)(t)$$

$$= x_1 \sum \alpha_1(t) + x_2 \sum \alpha_2(t) = x_1\sigma^{\alpha_1}(Q_{ij}) + x_2\sigma^{\alpha_2}(Q_{ij}),$$

so that the subset sum is a covariant of the vector space $A_F(T)$.

Let $S(T,Q)$ be the set of all α in $A_F(T)$ which satisfy the following semi-magic conditions:

(M.1)

> there exist numbers $\sigma_i(\alpha)$ for $i = 1, 2, \ldots, k$ such that
> $$\sigma^\alpha(Q_{ij}) = \sigma_i(\alpha) \qquad \text{for } j = 1, 2, \ldots, m_i.$$

That is, within each class of distinguished subsets there is a constant class sum.

The set $S(T,Q)$ is never empty, since the zero vector ζ satisfies (M.1) with every $\sigma_i(\zeta) = 0$. The set $S(T,Q)$ is a subspace of $A_F(T)$ which we call the semi-magic space of (T,Q). For if both α_1 and α_2 belong to $S(T,Q)$ so that

$$\sigma^{\alpha_1}(Q_{ij}) = \sigma_i(\alpha_1) \quad \text{and} \quad \sigma^{\alpha_2}(Q_{ij}) = \sigma_i(\alpha_2),$$

then $x_1\alpha_1 + x_2\alpha_2$ belongs to $S(T,Q)$ with

$$\sigma_i(x_1\alpha_1 + x_2\alpha_2) = x_1\sigma_i(\alpha_1) + x_2\sigma_i(\alpha_2),$$

so that the constant class sums are covariants of the vector space $S(T,Q)$.

Let $Z(T,Q)$ be the set of all α in $S(T,Q)$ such that

(M.2) $\qquad \sigma_i(\alpha) = 0$ for $i = 1, 2, \ldots, k.$

Note that $Z(T,Q)$ is a subspace of $S(T,Q)$ and call $Z(T,Q)$ the zero-magic space of (T,Q).

Thus we have a nest of vector spaces: $\zeta \subseteq Z(T,Q) \subseteq S(T,Q) \subseteq A_F(T)$. (Note the possibility of a lattice (by inclusion) of 2^k subspaces, topped by $S(T,Q)$ and bottomed by $Z(T,Q)$, when the condition $\sigma_i(\alpha) = 0$ is imposed for selected combinations of ν of the indices $0 \le \nu \le k$.) Using dimension as a criterion we define

$\quad (T,Q)$ is trivially magic if $\dim S(T,Q) = 0$;

$\quad (T,Q)$ is zero-magic $\qquad$ if $\dim S(T,Q) = \dim Z(T,Q) > 0$;

$\quad (T,Q)$ is semi-magic $\qquad$ if $\dim S(T,Q) > \dim Z(T,Q)$.

We say (T,Q) is pseudo-magic if there exists an α in $S(T,Q)$ such that

(M.3) for all pairs in T, if $t \ne t'$, then $\alpha(t) \ne \alpha(t')$.

Henceforth we assume F to be a subfield of the real field. We are happy to report that the necessary and sufficient conditions for a system to be pseudo-magic which were established for a special situation in Theorem 1 and Theorem 5 of [1] extend almost word for word to "magic sets".

We say (T,Q) is magic if there exists an α which is pseudo-magic and has

(M.4) $\qquad \alpha(t) > 0$ for all t in T.

We say (T,Q) is supermagic if there is an α in $S(T,Q)$ for which

(M.5) the set $\{\alpha(t_j)\}$ consists of consecutive positive integers.

We say (T,Q) is prime-magic if there exists an α in $S(T,Q)$ such that

(M.6) the set $\{\alpha(t_j)\}$ consists of distinct positive primes.

With this much of a general outline we now turn to specific examples.

In Stewart [1] and [2] the set T consists of the edges of an ordinary finite graph G with no isolated vertices; and the subsets of T in Q belong to just one class: for each vertex of G, consider the subset of T consisting of the edges of G which are incident to the vertex. Thus the "subset sum" is a "vertex sum" and condition (M.1) is a "constant vertex sum". For sample theorems we mention that if G is connected and has circuit rank C(G), then $C(G) \leq \dim S \leq 1 + C(G)$; also $\dim Z \leq \dim S \leq 1 + \dim Z$. A complete graph K_n is supermagic in this context if and only if $n = 2$ or if $n > 5$ and $n \not\equiv 0 \mod 4$. In [1] examples are given to show that the magic categories described above are all infinite, except possibly for the prime-magic case.

In Kotzig and Rosa [3] the set T consists of the vertices and edges of a finite graph G; and the subsets of T in Q belong to just one class: each distinguished subset consists of an edge and its two endvertices. These authors did not use the vector space description, but might easily have shown that $\dim S = 1 + |V|$. For sample theorems I mention the following: All n-gons are supermagic; and a complete graph K_n is supermagic in their context if and only if $n = 2, 3, 5$ or 6. They have a fascinating conjecture that every tree may be supermagic. They have shown this for all "two-rowed" trees; and I have shown it for all "two-length" and "three-length" stars.

In Domergue and Marsh [4] and Dongre [5] the set T consists of the vertices of a regular n-gon, $n \geq 5$, and the points of intersection of the chords of minimal length. The subsets belong to just one class: each subset consists of the four points on a chord. Various questions are asked but the principal one is whether (T,Q), the star-polygon, is supermagic. Thus far all that is known is that the cases $n = 6, 7$, and 9 are supermagic and that the case $n = 5$ is not. It was of great interest to me to discover that the problem in which T consists of the edges of the n-gon and the chords of minimal length, and in which each subset listed in Q is made up of the four edges incident to a vertex, is equivalent to the stated problem. I thought my greater familiarity with the vertex-sum problems would bring forth solutions as it did for the complete graphs; but I can discover no induction procedure for the star-polygon problem.

In Murty [6] the set T is a finite set of non-collinear points in E^2; and points of T belong to a distinguished subset if and only if they are on a line of E^2. The problem is to describe all sets T for which there is an α in $S(T,Q)$ satisfying (M.4), the "positiveness" condition. Only three types of configurations are known satisfying (M.1) and (M.4) and none of these is pseudo-magic. The problem is more geometric than magic, but the language of magic sets is somewhat convenient.

In conclusion let me describe a vector space way of deriving and remembering a "diabolic doughnut". On a torus a set of four latitudinal circles and four longitudinal circles creates an imbedding with 16 faces for a graph with 16 vertices and 32 edges; see Figure 1. Let T be the set of 16 faces. Let Q be the following set of 32 subsets of T, all in one class, and with each subset containing four faces: in a row (four rows); in a column (four columns); in a diagonal (eight diagonals); adjacent to a vertex (sixteen vertices).

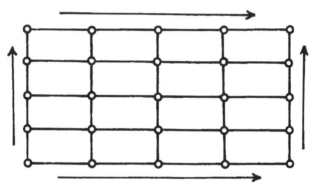

Figure 1.

Because of the imbedding on a torus, there is no need to treat broken diagonals and edge or corner vertices as special cases. An easy algebraic argument shows $\dim S(T,Q) = 5$. By inspection we can check that the following functions are in $S(T,Q)$ and form a linearly independent basis:

α_1:				α_2:				α_3:				α_4:				α_5:			
1	1	1	1	0	1	1	0	1	1	0	0	1	0	1	0	0	1	0	1
1	1	1	1	1	0	0	1	0	0	1	1	1	0	1	0	1	0	1	0
1	1	1	1	0	1	1	0	1	1	0	0	0	1	0	1	1	0	1	0
1	1	1	1,	1	0	0	1,	0	0	1	1,	0	1	0	1,	0	1	0	1.

In particular, $\alpha = \alpha_1 + \alpha_2 + 2\alpha_3 + 4\alpha_4 + 8\alpha_5$ is a supermagic solution, viz.,

$$
\begin{array}{cccc}
7 & 12 & 6 & 9 \\
14 & 1 & 15 & 4 \\
11 & 8 & 10 & 5 \\
2 & 13 & 3 & 16 \, .
\end{array}
$$

Note that the basis for $S(T,Q)$ given here is an "integral basis" (shades of algebraic number theory!), i.e., $\alpha = \Sigma \, x_i \alpha_i$ has $\alpha(t)$ a rational integer, for every t in T, if and only if every x_i is a rational integer.

References

1. B. M. Stewart, Magic graphs, Canad. J. Math. 18 (1966), 1031-59.

2. _____, Supermagic complete graphs, Canad. J. Math. 19 (1967), 427-38.

3. A. Kotzig and A. Rosa, Magic valuation of finite graphs, Canad. Math. Bull. 13 (1970), 451-61.

4. A. Domergue and D. C. B. Marsh, E2092, Magic number stars, Amer. Math. Monthly 76 (1969), 557-58.

5. N. M. Dongre, E2265, Magic star polygons, Amer. Math. Monthly 78 (1971), 1075.

6. U. S. R. Murty, How many magic configurations are there?, Amer. Math. Monthly 78 (1971), 1000-2.

Alan Tucker
SUNY at Stony Brook
Stony Brook, NY 11790

1. Introduction. The notation in this paper follows that of
Berge, see [3,4,5]. A graph G is called γ-perfect if λ(H) = γ(H)
for every vertex-generated subgraph H of G. Here, λ(H) is the
clique number of H (the size of the largest clique of H) and
γ(H) is the chromatic number of H (the minimum number of independent sets of vertices that cover all vertices of H). A graph G is
called α-perfect if α(H) = θ(H) for every vertex-generated subgraph
H of G, where α(H) is the stability number of H (the size of
the largest independent set of H) and θ(H) is the partition
number of H (the minimum number of cliques that cover all vertices
of H). For an arbitrary graph, we observe that λ(H) ≤ γ(H) and
α(H) ≤ θ(H). A graph is called perfect if it is both γ-perfect and
α-perfect. Berge [3,4,5] has examined perfect graphs extensively
(also see [9]). Many familiar classes of graphs, e.g., bipartite
graphs, are perfect. Moreover, α-perfect graphs and γ-perfect graphs
have great importance in several fields. Shannon [10] has shown that
graphs G such that α(G) = θ(G) represent perfect channels in
communication theory. Naturally, γ-perfection is important in
coloring problems, both in pure coloring problems and in applied ones
in block design or operations research. Later in this paper, we present an interesting application of coloring theory and perfect graphs
to a problem in refuse collection. Note that the equations
λ(G) = γ(G) and α(G) = θ(G) are of substantial interest in their
own right, since they are complementary equations involving the complementary concepts of clique and independent sets. That is, a clique
in G is an independent set in G^c, the complement of G, and so
λ(G) = α(G^c) and γ(G) = θ(G^c). Berge [3] has conjectured the following theorem.

STRONG PERFECT GRAPH CONJECTURE. The following conditions on a
graph G are equivalent:

a) G is perfect,

b) G is γ-perfect,

c) G is α-perfect, and

d) G and its complement G^c contain no odd-length primitive
    (chordless) circuits besides triangles (for short, no OPC's).

Omitting condition d), we have the (Weak) Perfect Graph Conjecture. Fulkerson [7] has generalized the ideas of α-perfection and
γ-perfection to anti-blocking pairs of polyhedra and has proved what
he calls the Pluperfect Graph Theorem. He posed the Perfect Graph
Conjecture in terms of pluperfect graphs. Lovász [8] extended
Fulkerson's work to obtain a proof of the Perfect Graph Conjecture.

2. Strong Perfect Graph Conjecture for planar graphs. Recently,
this author [11] proved that the Strong Perfect Graph Conjecture is
valid for planar graphs. Since the complement of a planar graph G
cannot contain an OPC of length ≥ 7, the Strong Perfect Graph
Conjecture reduces to:

THEOREM 1. The following conditions on a planar graph G are
equivalent:

b) G is γ-perfect,

c)   G is α-perfect, and

d') G contains no OPC.

It is easy to check that b) or c) implies d'). By Lovász's
result, it suffices to show that d') implies b) or c). We outline a
proof for both implications, or equivalently, a proof that for any
planar graph G, d') implies $\lambda(G) = \gamma(G)$ and $\alpha(G) = \theta(G)$ (if G
contains no OPC, then no subgraph H will either, and thus
$\lambda(H) = \gamma(H)$ and $\alpha(H) = \theta(H)$). Each implication is proved by induction. The induction step in showing d') implies $\lambda(G) = \gamma(G)$ uses
a case-by-case argument involving the possible values of $\lambda(G)$
$(\lambda(G) \leq 4)$ and the lowest degree of a vertex in G (known to be
≤ 5). While the proof becomes quite involved for $\lambda(G) = 3$, the
proof for $\lambda(G) = 4$, the "Four-Color Theorem", is quite easy given
condition d'). Let x be a vertex of lowest degree and assume G - x

has been 4-colored. If deg(x) ≤ 3, the result is trivial. We show
the argument for deg(x) = 5 (the case deg(x) = 4 is similiar). In
clockwise order let a, b, c, d, e be the vertices adjacent to x
and let 1, 2, 3, 4 be the colors. Assume all 4 colors are used in
coloring N(x) = a, b, c, d, e or else x can be colored immediately.
Figures 1a and 1b show the two different situations which can occur.
We can assume that no non-consecutive vertices in the set N(x) are
adjacent (if, say, b and d are adjacent, we disconnect c from
x,b and d; by induction, color the 2 components with c being
given a color different from x,b and d; and then re-connect c).
In Figure 1a, consider the subgraph generated by 2- and 4-vertices
and look at the component C of this subgraph containing c. If e
is in C, then C contains a (chordless) odd length path P from c
to e; but then P together with edges (e,x) and (x,c) yields an
OPC. Hence e ∉ C. Now we interchange colors 2 and 4 in C and
color x with 2. In Figure 1b, the same argument works with the
component containing b in the 2-4 subgraph.

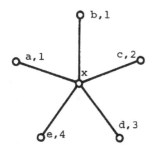

Figure 1a.

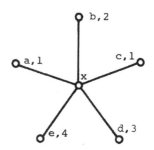

Figure 1b.

For the inductive proof of the second part of Theorem 1, we need
the following lemma (which may be of use in proving the general Strong
Perfect Graph Conjecture). Let N(x) denote the set of vertices adjacent to x (x ∉ N(x)). Let IS(G) denote the set of maximal independent sets of G. Then for I ∈ IS(G), let $I_x = I \cap N(x)$.

LEMMA 2. Let G be a graph with no OPC's. Let x be a vertex and let $I, I' \in IS(G)$ be such that a is not adjacent to b or $a = b$ for any $a \in I_x$, $b \in I'_x$. Then there exists $I'' \in IS(G)$ such that $I''_x = I_x \cap I'_x$.

Proof. Consider the $(I\&I')$-components of G, that is, the components of the $(I\&I')$-subgraph (generated by vertices of I and I'). A simple counting argument shows that we get a new pair of maximal independent sets by doing an $(I\&I')$-interchange in one or more, but not all, of the $(I\&I')$-components.

If $c \in I \cap I'$, then c must be an isolated point in the $(I\&I')$-subgraph. We claim that if an $(I\&I')$-component C with more than one vertex contains a vertex a in I_x, then C contains no vertices of I'_x. Suppose there exists $b \in C \cap I'_x$. Since $a \neq b$ (for otherwise C consists of only a), there exists a primitive $(I\&I')$-path P from a to b of odd length k ($k > 1$ since by assumption a is not adjacent to b). Assume b is the only vertex of I'_x on P (if not, pick a new b and shorten P). Similarly, we assume a is the only vertex of I_x on P. Then P plus edges (b, x) and (x, a) is an OPC, and the claim follows. Now in all non-isolated $(I\&I')$-components containing vertices of I_x, do an $(I\&I')$-interchange. This turns I into a new maximal independent set I'' with $I''_x = I_x \cap I'_x$.

COROLLARY 2.1. Let G be a graph with no OPC's such that all proper subgraphs are α-perfect. If G is not α-perfect, then for any vertex x of G, some pair of vertices in $N(x)$ is adjacent.

Proof. Suppose G is not α-perfect. Then removing a clique C of vertices from G will not lower α, i.e., $\alpha(G) = \alpha(G-C)$. Suppose no pair of vertices in $N(x)$ is adjacent. For each $a \in N(x)$, pick an $I^a \in IS(G-x-a) \subseteq IS(G)$ (inclusion follows from $\alpha(G) = \alpha(G-\{x,a\})$). Then by multiple application of Lemma 2 to the I^a's, we obtain $I \in IS(G-x)$ with

$$I_x = \bigcap_{a \in N(x)} I^a_x = \phi.$$

Then $I \cup x$ is an independent set of size $\alpha(G) + 1$.

The inductive proof that a planar graph with no OPC's is α-perfect involves a straightforward case-by-case argument using Lemma 2 and Lemma 3 (the proof is omitted).

LEMMA 3. Let G be a planar graph with no OPC's and such that G is not α-perfect but all proper subgraphs are α-perfect. Then for any vertex x with $\deg(x) \leq 5$ and for any $I \in IS(G)$, $|I_x| \geq 2$.

3. An application to refuse collection. Theorem 1 has the following useful corollary for coloring problems.

COROLLARY 1.1. If G and G^c contain no OPC's and G has no clique of size k + 1, then G can be k-colored.

Recently, this author was shown a routing problem whose solution required frequent testing for the k-colorability of a certain type of graph (in only one instance need an actual coloring be found). Since the associated integer program (which gives a k-coloring when it exists) takes a long time to solve, it was hoped that graph theory could provide a quicker sufficiency test which would not be too strong (would not eliminate too many graphs which are k-colorable). This appears to be the first non-trivial application of coloring theory to operations research; and Corollary 1.1 (if true for graphs in general) could provide the needed weak sufficiency test.

The k-colorable graphs arise in the following problem which was posed to the Urban Science Department at Stony Brook by the City of New York [1]. There is a set of sites S_i which must be serviced k_i times each week ($1 \leq k_i \leq 6$; in this case, the visit was to pick up garbage). One wishes to derive a minimal, or near minimal, set of (day-long) truck tours for a week such that each site is visited k_i times and, in addition, such that these tours can be partitioned among the six days of the week (Sunday is excluded) in a manner so that no site is visited twice on one day. Even if each site is visited just once a week, this is, in general, an extremely difficult problem (which is complicated in this case by the fact that the garbage trucks must return periodically to their dumpsites). The method proposed for attacking this multiple-visit problem was a modification of an algorithm due to Clarke and Wright [6]. This algorithm starts with an inefficient set of tours and successively tries to improve the set of tours. This method only gives a near-minimal set of tours. Further, it does not check to see that the derived tour set can be partitioned among the days of the week as required. This is where the

coloring problem arises. Given a set of tours, we form an associated
tour graph with one vertex for each tour and make two vertices adjacent if they correspond to two tours which visit a common site.
Partitioning the tours among the 6 days is equivalent to 6-coloring
the tour graph. To insure that the final set of tours has a tour
graph which can be 6-colored, we would check each successive improvement tried by the Clarke-Wright algorithm to see that the resulting
tour graph G_i does not have a clique of size 7 and that neither
G_i nor G_i^c have an OPC (although Theorem 1 is not yet proved for
general graphs, this test is being implemented; were this test to fail
in some routing problem, then at least operations research's loss
would be mathematics' gain).

In practice, an additional constraint is needed. We want to require that a site visited two or three times a week not only must have
each visit on a different day but must also have the visits spaced
evenly through the week. There are several ways to realize this constraint. One can manually alter the near-minimal tour set to insure
even spacing, or one can solve the routing problem separately for each
half (or third) of the week. In the extreme case, one assigns each
site a specific set of days on which it is to be visited and then the
simple routing problem is solved for each day separately. These and
other approaches are currently being investigated.

In closing, we note that once the garbage gets to the dumpsites
(situated along rivers) and is loaded on barges, then there is a whole
new (but similar) routing problem for the tours of the tugboats that
push groups of barges down to a landfill site on Staten Island. Because of special constraints in this case (one is that a tugboat push-
ing full barges must move with the tides), a different (randomized)
variation of the Clarke-Wright algorithm was used [2].

References

1. S. Altman, N. Bhagat, and L. Bodin, Extension of Clarke and Wright
   algorithm for routing garbage trucks, *Management Science*, to
   appear.

2. E. Beltrami, N. Bhagat, and L. Bodin, Refuse disposal in New York
   City: An analysis of barge dispatching, *Transportation Science*,
   to appear.

3. C. Berge, Farbung von Graphen, deren Samtliche bzw. deren
   ungerade Kreise starr sind, *Wissen. Ziet. der Martin-Luther Univ.*,
   Halle-Wittenberg, 1961, 114-115.

4. C. Berge, Some Classes of Perfect Graphs, Graph Theory and Theoretical Physics, Academic Press, New York, 1967.

5. _____, The rank of a family of sets and some applications to graph theory, Recent Progress in Combinatorics (ed. W. Tutte), Academic Press, New York, 1969, 237-251.

6. G. Clarke and J. Wright, Scheduling of vehicles from a central depot to a number of delivery points, Operations Research 12 (1964), 568-579.

7. D. R. Fulkerson, The Perfect Graph Conjecture and Pluperfect Graph Theorem, 2nd. Chapel Hill Conference on Combinatorial Mathematics and its Applications, Chapel Hill, 1969, 171-175.

8. L. Lovász, Normal hypergraphs and the Perfect Graph Conjecture, Discrete Mathematics, to appear.

9. H. Sachs, On the Berge conjecture concerning perfect graphs, Combinatorial Structures and Their Applications (eds. R. Guy et. al.), Gordon and Breach, New York, 1970, 377-384.

10. C. Shannon, The zero-error of a noisy channel, IRE Trans. Info. Thy. IT-3 (1956), 3.

11. A. Tucker, The Strong Perfect Graph Theorem for planar graphs, submitted for publication.

ON GRAPHICAL REGULAR REPRESENTATIONS OF $C_n \times Q$

Mark E. Watkins
Syracuse University
Syracuse, NY 13210

A finite simple graph X with vertex set V(X) and automorphism group A(X) is a __graphical__ __regular__ __representation__ (GRR) of the finite group G if (i) $G \cong A(X)$, and (ii) A(X) acts on V(X) as a regular permutation group; that is, given $u, v \in V(X)$, there exists a unique $\varphi \in A(X)$ such that $\varphi(u) = v$. A group will belong to __Class__ I if it admits a GRR.

With the letter G always denoting a finite group, the letter H will always denote a subset of G such that (i) the identity $e \notin H$, (ii) $h \in H \Rightarrow h^{-1} \in H$, and (iii) H generates G. The group G is in __Class__ II if for each such H, there exists a group automorphism $\varphi \in \mathrm{Aut}(G)$ such that $\varphi[H] = H$, but φ is not the identity of Aut(G). Class I and Class II are disjoint [8, Theorem 1], and every group known not to be in Class I has actually been shown to belong to Class II. Thus the non-existence of a GRR for a group G seems to have strong consequences for the action of Aut(G) upon sets which generate G.

Large families of finite groups as well as numerous particular groups have been classified in this manner. Let us summarize some of these results of Imrich, Nowitz, Sabidussi, and the author, [1] - [5] and [7] - [10]. We let C_n denote the cyclic group of order n.

1. Abelian groups. The elementary abelian 2-group $(C_2)^n$ is in Class I when n = 1 or $n \geq 5$. All other abelian groups are in Class II.

2. Generalized dicyclic groups. All these groups, which include all the generalized quaternion groups, are in Class II.

3. All non-abelian groups whose order is coprime to 6 are in Class I.

4. Alternating groups. The alternating group A_n is in Class I for $n \geq 5$, and in Class II for n = 3,4.

5. Semi-direct products of two cyclic groups. All of these which are neither abelian nor generalized dicyclic are in Class I

except for the dihedral groups D_3, D_4, D_5, and one group of order 16.

6. Cyclic extensions of groups in Class I. Let the non-abelian group G_1 be a cyclic extension of the group G in Class I. If the index $[G_1:G] \geq 5$ or if $|G| > 36$, then G_1 is in Class I. (Observe that 4 and 6 imply that the symmetric group S_n is in Class I for $n \geq 5$. Incidentally, so is S_4.)

7. Direct products of two groups in Class I. With the exception of $C_2 \times C_2$, whenever G_1 and G_2 are in Class I, so is $G_1 \times G_2$.

8. The group of order $2 \cdot 3^n$ generated by the elementary abelian 3-group $(C_3)^n$ together with an involution b such that $bxb = x^{-1}$ for all $x \in (C_3)^n$ is in Class II if $n = 1$ or 2 and Class I for $n \geq 3$.

Other results classifying specific small groups abound, but hopefully 1 - 8 above lend credence to the author's conjecture in [9] that there exists a positive integer N such that if a non-abelian group G is in Class II, then either G is a generalized dicyclic group or $|G| \leq N$. If the conjecture were true, one would have $N \geq 27$, since the non-abelian group of order 27 and exponent 3 is in Class II [5].

In the present paper we classify the direct products $C_n \times Q$ where Q is the 8-element quaternion group. The principal result is the following:

THEOREM. The group $C_n \times Q$ is in Class II for $n \leq 4$ and in Class I for $n \geq 5$.

The effect of this theorem is two-fold. On the one hand, if the aforementioned conjecture is true, then one has $N \geq 32$ by virtue of the group $C_4 \times Q$. On the other hand, this theorem adds credence to the conjecture for the following reasons. Indeed, the abelian groups of exponent > 2 are "very much" in Class II in that they admit the inverting automorphism. The generalized dicyclic groups are in a sense almost as good since they have been characterized ([3], [8]) by the property of being the only non-abelian groups G admitting a non-identity automorphism $\varphi \in Aut(G)$ such that for each $x \in G$, either $\varphi(x) = x$ or $\varphi(x) = x^{-1}$. Since Q is the smallest generalized dicyclic group and since in general the smaller a group in a family the more easily it had been demonstrated to be in Class II, one is inclined to suppose that the groups $C_n \times Q$ would be very suitable candidates for counter-examples to the conjecture. Insofar as they fail to do this, the conjecture appears the more valid.

The quaternion group Q is defined by

$$Q = \langle x, y \rangle; \quad x^2 = y^2 = (xy)^2. \tag{1}$$

We represent the cyclic group C_n by

$$C_n = \langle c \rangle; \quad c^n = e . \tag{2}$$

In addition to (1) and (2), $C_n \times Q$ is characterized by

$$cx = xc \quad \text{and} \quad cy = yc ,$$

to yield a group of order $8n$.

The group $C_2 \times Q$ is readily seen to be generalized dicyclic since it is generated by the abelian group $M = \langle c, x \rangle \cong C_2 \times C_4$ and the element $y \notin M$ satisfying

$$y^2 \in M, \quad y^4 = e, \quad ymy^{-1} = m^{-1} \quad \text{for all} \quad m \in M.$$

We shall next demonstrate formally that $C_4 \times Q$ is in Class II. The careful reader will observe that implicit in this proof is the treatment of $C_3 \times Q$ which, therefore, need not be done separately. This demonstration is entirely group-theoretical.

Let an arbitrary subset $H \subseteq C_4 \times Q$ be given which generates $C_4 \times Q$ and such that $e \notin H = H^{-1}$, according to our convention. With regard to the coset decomposition

$$C_4 \times Q = Q \cup cQ \cup c^2Q \cup c^3Q,$$

let $H_i = H \cap c^iQ$ $(i = 0, 1, 2, 3)$. Since H generates $C_4 \times Q$, we have $H_1 \neq \emptyset$, and if $q \in Q$, then

$$q \in H_0 \quad \text{if and only if} \quad q^{-1} \in H_0,$$
$$cq \in H_1 \quad \text{if and only if} \quad c^{-1}q^{-1} \in H_3,$$
$$c^2q \in H_2 \quad \text{if and only if} \quad c^2q^{-1} \in H_2.$$

One should bear in mind that Aut(Q) includes all permutations of the three 2-sets

(3) $$\left\{x,x^{-1}\right\}, \quad \left\{y,y^{-1}\right\}, \quad \left\{xy,xy^{-1}\right\}$$

setwise, each 2-set in (3) consisting of an element of order 4 and its inverse.

Let us first suppose that H_1 either includes or is disjoint from some 2-set of the form c times a pair in (3), and by the preceding paragraph, it may be assumed without loss of generality that the 2-set in question is $\{cx,cx^{-1}\}$. Then let $\omega \in \text{Aut}(C_4 \times Q)$ be defined by

$$c \longmapsto c^{-1}, \quad x \longmapsto x, \quad y \longmapsto y^{-1}.$$

Observe that φ then interchanges cx with $c^{-1}x$ and cx^{-1} with $c^{-1}x^{-1}$ while $\omega(g) = g$ or g^{-1} for all other $g \in C_4 \times Q$. Hence $\varphi[H] = H$.

Without loss of generality one may now assume:

$$cx, \ cy, \ cxy \in H; \quad \text{and} \quad cx^{-1}, \ cy^{-1}, \ cxy^{-1} \notin H.$$

A 2-set $\{q,q^{-1}\}$, where $q \in Q$, which is the only 2-set (3) to be contained in or disjoint from H_0 will be called the <u>distinguished</u> <u>pair</u> of H_0. The only 2-set $\{c^2q,c^2q^{-1}\}$ to be contained in or disjoint from H_2 is the <u>distinguished</u> <u>pair</u> of H_2, if there be any. Without loss of generality we may assume that if H_0 has a distinguished pair, then that pair is $\{x,x^{-1}\}$.

If the distinguished pair of H_2 is $\{c^2x,c^2x^{-1}\}$, or if H_2 has no distinguished pair, let $\omega \in \text{Aut}(C_4 \times Q)$ be defined by

$$c \longmapsto c^{-1}, \quad x \longmapsto x^{-1}, \quad y \longmapsto xy^{-1}.$$

Then $\varphi[H_0] = H_0$, $\varphi[H_2] = H_2$, and

$$\varphi(cx) = \varphi(c)\varphi(x) = c^{-1}x^{-1} = (cx)^{-1},$$

$$\varphi(cy) = \varphi(c)\varphi(y) = c^{-1}xy^{-1} = (cxy)^{-1},$$

$$\varphi(cxy) = \varphi(c)\varphi(x)\varphi(y) = c^{-1}x^{-1}xy^{-1} = (cy)^{-1}.$$

So $\varphi[H_1] = H_1^{-1} = H_3$, and so $\varphi[H] = H$.

Suppose, therefore, that the distinguished pair of H_2 is not $\{c^2x, c^2x^{-1}\}$, and again without loss of generality, one may assume it is $\{c^2y, c^2y^{-1}\}$. There now remain two subcases to consider. First suppose that H_0 and H_2 either both contain or both exclude their distinguished pairs. In this case, their distinguished pairs may be interchanged by the element $\varphi \in \mathrm{Aut}(C_4 \times Q)$ defined by

$$c \longmapsto c^{-1}, \quad x \longmapsto c^2y, \quad y \longmapsto c^2x .$$

If, however, precisely one of H_0 and H_2 contains its distinguished pair, define φ by

$$c \longmapsto c, \quad x \longmapsto x, \quad y \longmapsto c^2y^{-1} .$$

The verification is then straightforward that $\varphi[H] = H$. Of course, in none of these cases was φ the identity, and so $C_4 \times Q$ is in Class II.

We can complete the proof of the theorem by constructing a GRR X for $C_n \times Q$ when $n \geq 5$. G. Sabidussi has shown [6] that a necessary (but by no means sufficient) condition for a graph X to be a GRR of a group G is that it be a Cayley graph of G with respect to some subset $H \subseteq G$ as prescribed above. Thus $X_{G,H}$ has vertex set $V(X) = G$ and edge set:

$$E(X) = \{[g, gh] \mid g \in G, \ h \in H\}.$$

For $G = C_n \times Q$, we now fix the set H to be the subset

$$H = \left\{ x, x^{-1}, y, y^{-1}, cx, c^{-1}x^{-1}, cy, c^{-1}y^{-1}, cxy, c^{-1}xy^{-1}, c^2y, c^{-2}y^{-1} \right\}$$

and X to be the graph $X_{G,H}$.

Cayley graphs are always vertex-transitive, since $A(X)$ always admits the regular subgroup of all left multiplications $g \longmapsto g'g$ for g, $g' \in G$. It remains only to show that the stabilizer A_g of some vertex g, in particular, that $A_e = \{\varphi \in A(X) \mid \varphi(e) = e\}$, is trivial.

Let X_e denote the subgraph of X induced by the vertices adjacent to (but not including) e, that is, those identified with the elements of H. Clearly if $\varphi \in A_e$, then the restriction of φ to X_e belongs to $A(X_e)$. Moreover, L. Nowitz and the author have established [4, Prop. 2.3]:

LEMMA. Let $X_{G,H}$ be a Cayley graph. Let $K \subseteq H$. Suppose the stablizer $A_k = A_e$ for all $k \in K$. Then $A_k = A_e$ for all $k \in \langle K \rangle$, the subgroup of G generated by K.

For the Cayley graph at hand, the subgraph X_e is shown in Figure 1.

Let $\varphi \in A_e$ be arbitrarily chosen, and let φ_e be the restriction of φ to X_e. Since cx and $c^{-1}x^{-1}$ are the only vertices of X_e whose valence (in X_e) is 3, they must be fixed or interchanged by φ_e. However $c^{-1}x^{-1}$ is adjacent to a pendant vertex $c^{-2}y^{-1}$ while cx is not. Hence cx is fixed by φ. Now y is fixed by φ by virtue of being the only vertex of valence 2 adjacent to a vertex

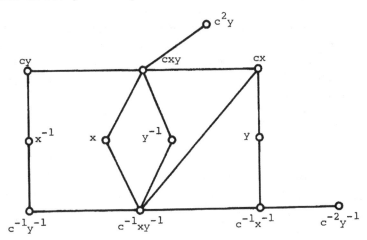

Figure 1.

of valence 3. Furthermore, x^{-1} is the only 2-valent vertex both of whose neighbors are 2-valent. Hence x^{-1}, y, cx are fixed points of the subgroup A_e. But $\{x^{-1}, y, cx\}$ generates all of $C_n \times Q$, and so by the Lemma, A_e consists of the identity alone. The proof is complete.

References

1. W. Imrich, Graphs with transitive Abelian automorphism group, _Combinatorial Theory and Its Applications_: Proceedings of the Combinatorial Conference at Balatonfüred, Hungary, 1969.

2. W. Imrich and M. E. Watkins, On graphical regular representations of cyclic extensions of groups, to appear.

3. L. A. Nowitz, On the non-existence of graphs with transitive generalized dicyclic groups, _J. Combinatorial Theory_ 4 (1968), 49-51.

4. L. A. Nowitz and M. E. Watkins, Graphical regular representations of non-abelian groups, I, _Canad. J. Math._, to appear.

5. _____, Graphical regular representations of non-abelian groups, II, _Canad. J. Math._, to appear.

6. G. Sabidussi, On a class of fixed-point free graphs, _Proc. Amer. Math. Soc._ 9 (1958), 800-804.

7. _____, On vertex-transitive graphs, _Monatsh. Math._ 68 (1964), 426-438.

8. M. E. Watkins, On the action of non-abelian groups on graphs, _J. Combinatorial Theory_ 11 (1971), 95-104.

9. _____, Graphical regular representations of alternating, symmetric, and miscellaneous small groups, to appear.

10. M. E. Watkins and L. A. Nowitz, On graphical regular representations of direct products of groups, _Monatsh. Math._, to appear.

PLANAR GRAPHS AND MATROIDS*

Louis Weinberg
The City College of New York
New York, NY 10031

1. Introduction. In preceding papers the author has presented
some results of his and his colleagues' research on planar graphs and
matroids. The present paper is based on these results and also ex-
tends them. For the most part we state the results and indicate how
other results have been obtained; for the proofs the reader should
consult the original papers.

Because of space limitations we assume the reader is familiar
with the basic definitions of graph theory or can consult the preceding papers. However, there are some definitions which are given
below to help in making the text immediately intelligible. Other
definitions appear where first used in the text.

We allow a graph to have no loops but multiple edges are allowed.
A graph that has no loops or multiple edges we call simple.

A graph G is said to be k-connected or k-tuply connected if it
contains at least k + 1 vertices and is not transformed to a disconnected graph by the deletion of less than k vertices. If G is
k-tuply connected but not (k+1)-tuply connected, then its connectivity
is defined as k.

Related to the concept of connectivity is that of "k-separators".
A graph G is k-separated if it is the union of two subgraphs H and
K with the following properties:

a.   H and K have no common edges, that is, they are
complements.

b.   H and K have not more than k common vertices.

c.   Each of the subgraphs H and K has a vertex not belonging
to the other.

We call the pair {H,K} a k-separator of G. A graph which is
not k-separated is called (k+1)-connected. Thus a graph that is not

* Research was partially supported by CUNY Research Foundation
Grant No. 1019.

2-separated is a triply connected graph. If {H,K} is a 2-separator
of G with two common vertices, we also speak of the common vertices
of H or of K.

Planar graphs constitute one of the most interesting subclasses
of graphs for which significant properties have been found. Some of
these properties give promise of leading to the solution of presently
unsolved problems for nonplanar graphs or give insights helpful in
determining such solutions. An example of such an unsolved problem
is the determination of an efficient algorithm for testing the isomorphism of two nonplanar graphs.

The concept of a matroid is a powerful generalization of that of
a graph. The theory of matroids is basic for the understanding of
fundamental graph-theoretic concepts like duality and the realizability of matrices as graphs. This is especially true for the concepts
of planar graphs, since by a theorem of Whitney that a graph is
planar if and only if it has a combinatorial dual, every characterization of matroids provides a characterization of planar graphs; conversely, a characterization of planar graphs, if it can be generalized
to a matroid, provides a realizability criterion for matroids.

First we give some alternative characterizations of planar graphs,
namely, three different sets of necessary and sufficient conditions
for a graph to be planar. This leads to a discussion of algorithmic
tests for planarity; and later in the second half of the paper, after
matroids are defined, we return to these characterizations. One of
the algorithms for planarity testing, where we use the concept of the
wheel, which was discovered by Tutte as a basic triply connected
graph, has been generalized to matroids.

We then derive codes for planar graphs of different connectivity;
specifically, we consider codes for triply connected graphs, and indicate that codes have also been formulated for trees and series-
parallel graphs. Since the codes form a one-to-one correspondence
with graphs, they are exceedingly useful in computer analysis of
systems. One use that we discuss is the determination of the isomorphism of planar graphs and their automorphism groups. It is also
important to note that isomorphism tests have recently been formulated
as automata.

Finally we consider matroids. After giving the circuits definition of a matroid, we refer to different characterizations of graphi
and cographic matroids due to Tutte and Welsh and indicate how they
reduce to the characterizations of planar graphs given respectively by
Kuratowski and MacLane. It should be unnecessary to point out that

the theorems on graphs were formulated before the corresponding matroid theorems. This of course is to be expected when one recognizes that the particularization of a matroid realizability theorem to planar graphs is _automatic_ and thus does not constitute a new result. At present the author and his doctorate students are carrying out research on the process of generalizing known theorems on planar graphs to theorems on matroid realizability. We have already done this in unpublished work for the characterization of planar graphs in terms of wheels and are now investigating other characterizations.

2. _Algorithms_ _for_ _planarity_ _testing_. This section is devoted mainly to a discussion of an algorithm that uses the concept of the wheel for testing whether a given triply-connected graph is planar (see [2], [6]). We also briefly introduce two other algorithms which depend on more widely known concepts, namely, the spanning tree (which we refer to as a tree) and the _star_ _cut-set_, where by the latter term we mean the cut-set consisting of all the edges incident to the same vertex. (If the vertex is a cut-vertex, the edges constitute a _cut_.) Each of these algorithms in effect constitutes a new set of necessary and sufficient conditions for a graph to be planar.

A _planar_ _graph_ is a graph that can be drawn in the plane without crossed edges; such a drawing is called a _plane_ graph. The edges of a plane connected graph form a set of adjoining polygons which divide the plane into regions called _faces_; thus a face is a region bounded by a circuit, where the region contains no edges. This particular type of a circuit is designated as a _mesh_. Thus a mesh is a set of edges that encircle the corresponding face. Frequently we order the edges of a mesh so that the sequence forms a closed walk that encircles the corresponding face in a specified sense, say, clockwise.

Since our algorithm that uses a tree as a starting point works with subgraphs that have cut-vertices, for the purposes of this algorithm we allow the region enclosed by a mesh to contain a subgraph connected to the mesh at a cut-vertex.

There are F meshes (or faces) including the _outer_ _mesh_ (or corresponding _infinite_ _face_) in a plane graph; the number of meshes is related to the number of edges b and the number of vertices n in a connected plane graph by Euler's formula

$$F + n - b = 2.$$

It is clear that any plane graph has a unique set of meshes, but there are in general many plane representations for a planar graph. This is not true, however, when the graph is triply connected: such a graph has a unique plane representation, that is, _it has a unique family of meshes_. For a graph of connectivity 2 all representations, that is, all the different sets of meshes, may be obtained by permutation of parallel subgraphs at the two common vertices of each 2-separator or by rotation of one subgraph that is connected to these two vertices.

It is intuitively obvious (and may be proved by use of the Jordan curve theorem) that an edge bc may be added between vertices b and c of a plane graph without crossing other edges if and only if b and c are on a common mesh. Thus we reach the important conclusion that if a plane graph G is triply connected and b and c are not on a common mesh, the addition of edge bc to G yields a nonplanar graph. However, if G is not triply connected, G may be redrawn to have a different family of meshes, and the vertices b and c may be made to lie on a common mesh by permuting any subgraphs that are in parallel at the common vertices of a 2-separator, or by rotating a subgraph connected to the rest of the graph at only these two vertices, or by rotating a subgraph at a cut-vertex.

A tree is clearly a planar graph, and a plane representation is found easily. A series-parallel graph is also planar; a plane representation may be found by reversing the process for testing whether a graph is series-parallel, since in any plane graph each edge may be replaced by a parallel set of edges or by a series set of edges without destroying planarity.

Thus the general algorithms need not be applied to trees or series-parallel graphs. In addition, it is clearly not necessary to consider disconnected graphs since a disconnected graph is planar if and only if its connected components are planar. If the connected graph is separable, we may divide it into its blocks, and, according to a theorem of Whitney [9], the graph is planar if and only if its blocks are planar. Of course, if one of the components is a tree it is not necessary to divide it into its blocks; the preliminary testing procedure should therefore detect trees and keep them intact as a single component. Thus the preceding establishes that only biconnected components need be considered. Furthermore, if the biconnected graph is a series-parallel graph, it may be drawn without use of any of the basic algorithms.

Moreover, there is a structural characterization of planar graphs due to MacLane [4] which shows that we need consider only simple triply connected graphs. If the graph G is not triply connected, we find a 2-separator {H,K} of G. We then connect an edge between the common vertices of H and do the same for K. We then repeat the step to find a 2-separator on each of these two augmented 2-separator components. This process is repeated as often as necessary until we obtain triply connected components and some trivial biconnected components. The theorem given by MacLane shows that the original graph is planar if and only if its triply connected components are planar. Finally, if we eliminate all but one edge of each set of parallel edges in a triply connected component, we see that we need test only simple triply connected graphs.

Though the above simplification is available to the computer programmer, we present our first two algorithms so that they apply to any connected graph, not merely to a triply connected graph. However, the algorithm that uses the wheel requires that we start with a simple triply connected graph.

Before considering the first algorithm, we would like to suggest why new sets of conditions on planarity are desirable. We therefore remind the reader of some well-known sets of conditions.

There are three different well-known characterizations of planar graphs. The first is due to Kuratowski, the second to Whitney, and the third to MacLane. They are:

1. A graph is planar if and only if it does not contain as a subgraph either of the two basic nonplanar graphs -- namely, K_5, the complete graph on five vertices, and the complete bipartite graph $K_{3,3}$ -- or homeomorphs of these two graphs. (By a homeomorph is meant a graph in which the operation of replacing an edge by a series connection of edges is performed one or more times.)

2. A graph is planar if and only if it has a combinatorial dual.

3. A graph is planar if and only if it contains a complete set of circuits such that no edge appears in more than two of these circuits. (By a complete set of circuits we mean a basis of the vector space of circuits modulo 2; in other words, every circuit in the graph can be expressed uniquely as a sum mod 2 of some of the circuits in the complete set.)

To use any of the above as the basis for an algorithm introduces combinatorial difficulties. The procedure using the wheel that we shall discuss leads to no such difficulties and is conceptually simple. The procedures using the tree and the star cut-set, however, require that the biconnected components of subgraphs of the original graph be determined. We consider these algorithms first since they are exceedingly simple and in addition demonstrate the advantage of insuring that all subgraphs that must be considered in the algorithm are triply connected.

The first algorithm that is now considered makes use of the tree as its starting point. The edges that are not in the tree we denote as links. Each link forms a fundamental circuit with the tree.

Suppose we are given a connected graph G. We number the vertices of G arbitrarily $1, 2, \ldots, n$. We now draw any spanning tree of G and label the links arbitrarily $L_1, L_2, \ldots, L_{F-1}$. Each link is of course defined by its two vertices, called its ends; thus $L_1 = bg$, $L_2 = cf, \ldots$, where $b, g, c, f, \ldots$ are vertex numbers. We now repeat a basic step that draws a set of subgraphs $G_1, G_2, \ldots, G_{F-1} = G$, where G_1 consists of the tree augmented by link L_1, G_2 consists of G_1 augmented by link L_2, and in general G_r consists of G_{r-1} augmented by L_r.

When we add L_1, a fundamental circuit is formed. We label this circuit as mesh M_1 and the outer mesh as M_2. We then label each vertex with the meshes on which it lies. If a vertex is on a 1-separator component and is not part of a mesh, it is labeled with the meshes of the vertex that is both a cut-vertex of this 1-separator and is part of a mesh. Now when we add L_2, another mesh is formed which is labeled M_3. If the vertices of L_2 lie on M_1, we have a choice of placing L_2 in M_1 or M_2. Whenever we have such a choice, we may choose the mesh arbitrarily since the choice makes no difference in the successful working of the algorithm.

We can now state the theorem on which the algorithm is based (see [6]).

THEOREM 1. A graph G is planar if and only if each of its subgraphs G_r, where $r = 1, 2, \ldots, F-2$, is a plane graph and the two vertices of L_{r+1}, the next link to be added, lie on a common mesh or can be made to lie on a common mesh by rotation of a subgraph about a cut-vertex or about the common vertices of a 2-separator, or by permutation of subgraphs connected in parallel between the two common vertices of a 2-separator of G_r.

Proof. The theorem is easily proved. To show necessity we assume G is planar. Now suppose that the condition does not hold for a subgraph G_r. Since all the possible families of meshes of G_r can be generated by the rotation and permutation operations specified in the theorem, we cannot find a plane representation such that the vertices of L_{r+1} lie on a common mesh. Thus L_{r+1} cannot be added without crossing an edge of every plane representation of G_r, and G_{r+1} is nonplanar. This contradicts the assumption that G is planar.

For showing sufficiency we assume the condition is true. Thus at each step we can add a link without causing a crossover. We therefore obtain the plane graph G. This completes the proof.

We now consider the second algorithm, which is based on cut-sets rather than circuits. We use (n-1) star cut-sets. Precisely the same principles of rotation of 1-connected subgraphs and rotation or permutation of parallel subgraphs apply to this case. The proof of the corresponding theorem also follows easily. After numbering the vertices of the given graph 1, 2, ..., n we consider the star cut-sets in order and draw them in the plane. In considering the star cut-set at vertex r, we need only add its incident edges that join r to any vertex k, where k > r. Such edges can, of course, be added if and only if the vertices to be joined have a common mesh or can be made to have a common mesh by the operations discussed previously.

The theorem can be stated as follows, where we use similar notation as before and designate a star cut-set at vertex r by S_r. In addition, we use a slightly different definition of G_r; it is the subgraph obtained after all the edges of S_r and some (but not all) of the edges of S_{r+1} have been added to G_{r-1}. (When all edges of S_{r+1} have been added, G_r becomes G_{r+1}.)

THEOREM 2. A graph G _is planar if and only if_ each of the subgraphs G_r, _where_ r = 1, 2, ..., n-2, _is a plane graph and the two vertices of each of the edges of_ S_{r+1} _that must be added lie on a common mesh or can be made to lie on a common mesh by rotation of a subgraph about a cut-vertex or about the two common vertices of a 2-separator, or by permutation of the subgraphs connected in parallel between the two common vertices of a 2-separator of_ G_r.

It should be clear that the number of operations required in the algorithms depends critically on the number of subgraphs that are not triply connected and that may therefore require testing for their 2-separators. It is therefore desirable to make the subgraph drawn in the plane a triply connected graph as early as possible. The question therefore arises as to how this can be accomplished. Still better, we would prefer to work only with triply connected subgraphs, if this is at all possible.

It is possible, as shown in the third algorithm that we now discuss. The graph denoted as a "wheel" that is used in this algorithm is just what the name implies: in effect, it has a "rim", a "hub", and "spokes". The _wheel_ W_n has n edges (n ≥ 3) in its outer mesh, with each vertex in this mesh joined to a central vertex by an edge; thus W_n contains 2n edges and n + 1 vertices. The wheel W_3 is the complete graph K_4.

It is easy to prove that every wheel is a simple triply connected graph. There is, moreover, an additional property of a wheel which is extremely important.

To explain this property we must first define two operations on graphs. Let G be a graph and let $e \in E(G)$. By "G op e" we mean the subgraph of G with edge set $(E(G) - \{e\})$ and vertex set $V(G)$; in other words, we delete edge e from G. By "G cl e" we mean the graph obtained from G by the following steps:

   i. Form G op e.

   ii. If v_1 and v_2 are the ends of e in G, then we coalesce these vertices in the graph G op e and denote the resulting graph by G cl e.

In other words, G cl e denotes the new graph obtained by contracting e to a vertex.

Now we can define the property of wheels used in the planarity algorithm. Let G be a simple, triply connected graph. We call an edge $e \in E(G)$ an _essential_ _edge_ if neither G op e nor G cl e is both simple and triply connected. It is easy to see that _in a wheel every edge is essential_. Tutte has proved the converse statement: if G is a simple, triply connected graph in which every edge is essential, then G is a wheel.

The heart of the algorithm is a reduction sequence. Let G be a simple triply connected graph. (This, it should now be clear, does not restrict the generality of the procedure since if G is not simple and triply connected, we can decompose it into simple, triply connected graphs so that, by the theorem of MacLane previously mentioned, G is planar if and only if its associated simple triply connected graphs are planar.) Then the sequence of graphs

$$G_0, \ G_1, \ G_2, \ \ldots, \ G_r$$

is called a <u>reduction sequence</u> of G if the following conditions are satisfied:

    i. G_k is a simple triply connected graph for $k = 1, \ldots, r$;

    ii. $G_0 = G$;

    iii. G_r is a wheel; and

    iv.   either $G_{k+1} = G_k$ cl e   or $G_{k+1} = G_k$ op e, where $e \in E(G_k)$ for $k = 0, \ldots, r-1$.

Reduction sequences are easily constructed. Since G_r is a wheel, that is, every edge is essential, we merely test the graphs in the reduction sequence for non-essential edges. Having found one, we remove it to obtain a smaller graph. The process is repeated until there are no more non-essential edges and thus the process has yielded a wheel G_r.

Now we reverse the process that yielded the reduction sequence. Starting with the plane graph G_r, we use the appropriate inverse operation $(op)^{-1}$ or $(cl)^{-1}$ to reinsert edges to obtain G_{r-1}, G_{r-2}, etc., until a nonplanar graph is obtained or until we have a drawing of G as a plane graph.

The test to determine if the reinsertion of an edge yields a non-planar graph is simple in that it is checked by inspection. It is based on Whitney's theorem that the family of meshes of a planar triply connected graph are unique. Thus <u>an edge</u> e <u>can be reinserted in the operation</u>

$$G_k = G_{k+1} (op)^{-1} e$$

<u>to yield a planar graph if and only if the ends of</u> e <u>in</u> G_{k+1} <u>both lie in the same mesh.</u>

The condition for reinserting an edge that has been removed by the cl operation is the dual of the above. Thus if v_L and v_R are the ends of e that are coalesced into v, and L is the set of edges not including e that are incident to v_L, and R is the set incident to v_R, then e can be reinserted by the operation

$$G_k = G_{k+1}(c1)^{-1}e$$

to yield a planar graph if and only if the edges in L are located consecutively around the coalesced vertex v, that is, there exists an edge ℓ_1 in L such that if one moves in a clockwise direction around v beginning with ℓ_1, then all the edges of L are encountered before reaching a member of R.

It should be clear that the reverse process can be easily carried out. A program has been written and executed successfully on a computer.

3. Codes, isomorphism, automorphism groups. We wish to determine codes for planar graphs, that is, a one-to-one correspondence between a family of graphs and a family of codes (see [7]). We call such a code for a graph a "canonic code". These codes, in addition to allowing us to manipulate graphs in computers, solve the isomorphism problem and the automorphism group problem.

We will consider planar triply connected graphs first and show how a code may be generated by the stratagem of traversing a maze. We then briefly mention how such codes and others may be found for trees and series-parallel graphs. In addition, we suggest how we have derived even simpler codes based on the facts that the meshes (and star cut-sets) of a planar triply connected graph are unique and that the maximum order of the automorphism group of such a graph is equal to 4b, where b is the number of edges.

Assume that we are given a plane triply connected graph; that is, the graph is drawn in the plane without crossed edges, or, equivalently, it is stored in the computer (with the edges incident to each vertex listed in a counterclockwise order) so that when represented in the plane it contains no crossed edges.

The algorithm is based on the traversal of an Euler circuit. It is well known that Euler was concerned with the problem of finding a circuit (along an undirected connected graph) that traverses each edge of the graph once and only once. He showed that a necessary and sufficient condition for the existence of such a circuit is that the degree of each vertex be even. Euler, however, did not give a procedure for finding such a circuit. The one we use was formulated by Trémaux (see [1],p.255). To Tremaux's procedure we add a numbering scheme so

that a code is generated. To guarantee that a given graph will yield
an Euler circuit, we traverse each edge precisely once in <u>each</u> direction; thus we have effectively doubled the number of edges, which
guarantees that the degree at each vertex is even. In effect, we now
have a digrected graph that satisfies the condition for a directed
graph to have an Euler circuit -- namely, that the inward degree is
equal to the outward degree at each vertex.

In carrying out the algorithm we choose any edge in the plane
graph and a direction on that edge as an initial edge and direction;
this choice fixes the right mesh of that directed edge. Since there
are b edges, there are 2b different choices of an initial edge
and direction. Using the mirror-image graph gives an additional 2b
choices in which the right and left meshes of each initial edge reverse their roles. Thus we get a total of 4b choices.

It will become clear from the algorithm that the code corresponding to the choice of initial edge, edge direction, and right
mesh of the edge is unique. Thus using this procedure on each of the
4b choices permits us to obtain the maximum possible number of 4b
different codes, as specified by the theorem cited on the maximum
order of the automorphism group. As a consequence the code matrix defined later in terms of these 4b codes is unique for a triply connected graph.

The rules for the algorithm are given in the following. We call
a vertex <u>old</u> or <u>new</u> if it has or has not been reached previously,
respectively. An edge being traversed in one direction is termed <u>old</u>
if it has been previously traversed in the opposite direction, and
<u>new</u> otherwise. In traversing an edge we go from an initial vertex to
a terminal vertex. Starting with an edge traversed in one of its
directions, we follow these rules:

1. When a new vertex is reached, take (that is, leave the vertex
on) the right-most edge of the edge on which the vertex is reached.
For the list of incident edges stored in a computer, this corresponds
to choosing the edge that immediately follows the edge being traversed.

2. When an old vertex is reached on a new edge, go back in the
opposite direction on this edge.

3. When an old vertex is reached on an old edge, leave the vertex on the right-most edge that has not previously been traversed in
that direction (that is, with its initial vertex being the vertex just
reached). Thus edges are traversed only once in each direction.

4. We record each traversed edge and the direction of traversal.

5. As we visit (or reach) a new vertex we label it with a number in natural order. Thus the initial vertex of the first edge is 1, its terminal vertex is 2, and the next new vertex is labeled 3.

6. We form a code consisting of the numbers of the vertices visited in the order in which they are visited. The code is thus given by a vector of dimension $2b + 1$.

After performing this process with each of the b edges as an initial edge and using both directions on each initial edge, we have a set of $2b$ vectors. We now repeat the above on the mirror-image graph so that an additional $2b$ vector codes are obtained. We then order the $4b$ codes lexicographically (or equivalently, in order of increasing magnitude, where the codes are considered as numbers written to a base $n + 1$ if the number of vertices is $n \geq 10$) to obtain a unique code matrix. We designate this code matrix by M; thus the order of M is $4b \times (2b + 1)$.

As stated above, a code is specified by an ordered set of $2b + 1$ numbers $X = X_1 X_2 \cdots X_{2b+1}$, and thus may be represented by a vector $X = (X_1, X_2, \ldots, X_{2b+1})$. The numbers used in the vector are the integers $1, 2, \ldots, n$, where n is the number of vertices in the graph.

If we are given a code X, we can, by reversing the procedure for generating the code, draw a plane representation of the graph; this graph may have a different infinite face, which of course is immaterial, but will have the same sense of the ordering of the edges around each face.

It is clear that a code has a number of distinguishing properties. We can, in fact, specify a set of necessary and sufficient conditions by which a vector may be checked to see whether it qualifies as a code. These are given for a simple graph by the following theorem.

THEOREM 3. Let G be a plane triply connected graph that is simple. A vector $X = (X_1, X_2, \ldots, X_{2b+1})$ is a code for G if and only if X satisfies the following conditions:

1. The dimension of X is $2b + 1$, where b is the number of edges in G; the numbers used in the code are $1, 2, \ldots, n$, where n is the number of vertices in G.

2. If numbers occurring for the first time in the code are ordered in a sequence in accordance with their position in the code, reading from left to right, then they must be in natural order 1, 2, ..., n.

3. There are b different ordered pairs of successive numbers X_iX_j in the code, where $X_i \neq X_j$, and corresponding to each of these there is a pair in the reverse order X_jX_i.

4. If the ordered pair of successive numbers X_iX_j occurs in the code and the number X_j has appeared previously but X_jX_i has not yet occurred, then the next number must be X_i; thus we obtain the triplet $X_iX_jX_i$. If X_jX_i has occurred previously or X_j has not appeared previously, the next number is $X_k \neq X_i$ so that we obtain $X_iX_jX_k$.

5. For the first occurrence of the number X_j in the code, where $X_j = 2, 3, ..., n$, we note the immediately preceding number X_i. Now find X_jX_i in the code; then X_j cannot appear again in the code after this appearance of X_jX_i.

Now we return to the consideration of the code matrix M. We choose the first code as the code for the graph; this is the canonic code. Given a graph it is clear that its canonic code is unique, and given a canonic code, it is clear that it corresponds to a unique graph.

The first g codes of the code matrix M are equal, where $1 \leq g \leq 4b$, and there are $r = 4b/g$ different sets of equal codes. If $g = 1$, then the automorphism group of the graph is the identity. In general, g gives the order of the automorphism group. In fact, complete information about isomorphism of two graphs and the automorphism group of a graph is given by the code matrices, as stated in the following theorems.

THEOREM 4. Let G_1 and G_2 be two triply connected plane graphs and M_1 and M_2 their respective code matrices. The graphs are isomorphic if and only if their code matrices are equal.

Clearly, we need not use the complete matrices for determining isomorphism. For example, we have the following theorem.

THEOREM 5. Let G_1 and G_2 be two plane triply connected graphs and M_1 and M_2 their respective code matrices. Then G_1 and G_2 are isomorphic if and only if either of the two equivalent conditions is true:

1. The first row of M_1 equals the frist row of M_2.

2. Any row of M_1 equals any row of M_2.

The information about the automorphism group of a graph is given by the following theorem.

THEOREM 6. Let M be the code matrix for a triply connected plane graph G. Consider the first set of equal codes of M and let the original vertex numbering correspond to the first code. Then the elements of the automorphism group of G are given by the permutations of the vertex numbering corresponding to each of the codes in the first set of equal codes.

We conclude this discussion of the Euler codes by pointing out that it is possible to represent this maze-solving algorithm as an automaton. A comparable representation as an automaton of an iso-morphism-testing algorithm has been given by Hopcroft [3].

For similar codes for trees and series-parallel graphs and for other codes of triply connected graphs, the reader is referred to [7]. The basic problem of obtaining a unique representation in the plane for the tree is solved by use of the concepts of the center and central edge of a tree, and unique codes are obtained both for the tree and the series-parallel graph by use of Polish parenthesis-free notation. New codes for triply connected graphs are based on the uniqueness of the family of meshes and on an ordering algorithm [7].

4. Matroids and graphs. We now briefly consider "matroids" to determine how they add to our understanding of planar graphs, and conversely.

First we define a matroid and then consider it as a generaliza-tion of a graph.

Let E be the finite set $\{e_1, e_2, \ldots, e_n\}$. Let C be a class of non-null subsets of E called circuits. The pair $m = (C, E)$ is called a matroid on E if C satisfies the following two axioms:

C1. No member of C is contained in another member (that is, no element of C is a proper subset of any element of C).

C2. If C_1 and C_2 are members of C and e_1 and e_2 are members of E such that $e_1 \in (C_1 \cap C_2)$ and $e_2 \in (C_1 - C_2)$, then there exists a $C_3 \in C$ such that

$$e_2 \in C_3 \subseteq [(C_1 \cup C_2) - \{e_1\}].$$

The above two axioms are true for circuits and cut-sets of graphs, which are called <u>polygons</u> and <u>bonds</u>, respectively by matroid theorists; we shall now use the latter terms. Thus a matroid may be looked upon as a generalization of a graph. We make this notation precise below.

Let G be a graph with the set of edges $E(G)$. We associate a unique matroid to G as follows. Let E of the matroid be the set of edges $E(G)$. Let C of the matroid be the class of polygons of G denoted by $P(G)$. The resulting matroid

$$M(G) = (P(G), E(G))$$

is called the <u>polygon matroid</u> of G.

Another unique matroid may be associated to G. Let $B(G)$ be the class of bonds of G. Then

$$M(G) = (B(G), E(G))$$

is a matroid of the graph G. It is called the <u>bond matroid</u> of G.

Thus every graph has two unique matroids associated with it. However, if we are given a matroid, it may be impossible to find an associated graph. If a matroid is realizable as the polygon matroid of a graph -- that is, there exists a graph whose polygon matroid is the given matroid -- we say it is <u>graphic</u>. If it is realizable as the bond matroid of a graph, it is <u>cographic</u>. Clearly not all matroids are realizable as graphs; for example, the matroid on four elements, where C is the family of all the three-element subsets, is neither graphic nor cographic.

By the theorem of Whitney that a graph is planar if and only if it has a dual, every realizability criterion for matroids gives a theorem on planar graphs. Thus we call a matroid a <u>planar matroid</u> if it is both graphic and cographic.

Tutte [5] and Welsh [8] have generalized theorems on planar
graphs to realizability conditions on matroids. Tutte's theorem reduces to Kuratowski's theorem on planar graphs and Welsh's theorem
reduces to MacLane's theorem.

At present we are trying to generalize other known conditions for
planarity. The author and his doctorate student T. Inukai have accomplished this for the characterization of a planar graph in terms of
a wheel. For matroids the generalization requires the use of not only
the wheel matroid, but also a whirl matroid and a binomial coefficient
matroid. This result should be published soon.

5. <u>Conclusion</u>. We have discussed planar graphs, their characterization, testing and generation of codes. We have shown that the
codes solve the isomorphism and automorphism group problems. Since
matroids give a complete theory of duality for graphs, it is to be
expected that they yield insights and characterizations for planar
graphs. Conversely, a fertile field of research is to generalize
known conditions on planar graphs to realizability conditions on
matroids. Tutte has provided one such theorem and Welsh another. The
wheel algorithm for testing planarity that we discussed yields a third
characterization of matroids.

References

1. W. W. R. Ball, <u>Mathematical Recreations and Essays</u>, Macmillan,
   New York, 1962.

2. J. Bruno, K. Steiglitz, and L. Weinberg, A new planarity test
   based on 3-connectivity, <u>IEEE Trans. Circuit Theory</u> CT-17 (1970),
   197-206.

3. J. Hopcroft, An n log n algorithm for isomorphism of planar triply
   connected graphs, <u>Technical Report</u> CS-192, Stanford University,
   Stanford, 1970.

4. S. MacLane, A structural characterization of planar combinatorial
   graphs, <u>Duke Math. J.</u> 3 (1937), 460-472.

5. W. T. Tutte, Matroids and graphs, <u>Trans. Amer. Math. Soc.</u> 90
   (1959), 527-552.

6. L. Weinberg, Two new characterizations of planar graphs,
   <u>Proceedings of the Fifth Annual Allerton Conference on Circuit
   and Systems Theory</u>, University of Illinois, 1967.

7. L. Weinberg, Linear graphs - theorems, algorithms, and applications, Aspects of Network and Systems Theory, (edited by N. de Claris and R. Kalman), Holt, Rinehart, and Winston, New York, 1971, 61-162.

8. D. J. A. Welsh, On the hyperplanes of a matroid, _Proc. Camb. Phil. Soc._ 65 (1969), 11-17.

9. H. Whitney, Non-separable and planar graphs, _Trans. Amer. Math. Soc._ 34 (1932), 339-362.

Lecture Notes in Mathematics

Comprehensive leaflet on request

Please turn over

Vol. 178: Th. Bröcker und T. tom Dieck, Kobordismentheorie. XVI, 191 Seiten. 1970. DM 18,–

Vol. 179: Séminaire Bourbaki – vol. 1968/69. Exposés 347-363. IV. 295 pages. 1971. DM 22,–

Vol. 180: Séminaire Bourbaki – vol. 1969/70. Exposés 364-381. IV, 310 pages. 1971. DM 22,–

Vol. 181: F. DeMeyer and E. Ingraham, Separable Algebras over Commutative Rings. V, 157 pages. 1971. DM 16.–

Vol. 182: L. D. Baumert. Cyclic Difference Sets. VI, 166 pages. 1971. DM 16,–

Vol. 183: Analytic Theory of Differential Equations. Edited by P. F. Hsieh and A. W. J. Stoddart. VI, 225 pages. 1971. DM 20,–

Vol. 184: Symposium on Several Complex Variables, Park City, Utah, 1970. Edited by R. M. Brooks. V, 234 pages. 1971. DM 20,–

Vol. 185: Several Complex Variables II, Maryland 1970. Edited by J. Horváth. III, 287 pages. 1971. DM 24,–

Vol. 186: Recent Trends in Graph Theory. Edited by M. Capobianco/ J. B. Frechen/M. Krolik. VI, 219 pages. 1971. DM 18.–

Vol. 187: H. S. Shapiro, Topics in Approximation Theory. VIII, 275 pages. 1971. DM 22,–

Vol. 188: Symposium on Semantics of Algorithmic Languages. Edited by E. Engeler. VI, 372 pages. 1971. DM 26,–

Vol. 189: A. Weil, Dirichlet Series and Automorphic Forms. V. 164 pages. 1971. DM 16,–

Vol. 190: Martingales. A Report on a Meeting at Oberwolfach, May 17-23, 1970. Edited by H. Dinges. V, 75 pages. 1971. DM 16,–

Vol. 191: Séminaire de Probabilités V. Edited by P. A. Meyer. IV, 372 pages. 1971. DM 26,–

Vol. 192: Proceedings of Liverpool Singularities – Symposium I. Edited by C. T. C. Wall. V, 319 pages. 1971. DM 24,–

Vol. 193: Symposium on the Theory of Numerical Analysis. Edited by J. Ll. Morris. VI, 152 pages. 1971. DM 16,–

Vol. 194: M. Berger, P. Gauduchon et E. Mazet. Le Spectre d'une Variété Riemannienne. VII, 251 pages. 1971. DM 22,–

Vol. 195: Reports of the Midwest Category Seminar V. Edited by J.W. Gray and S. Mac Lane.III, 255 pages. 1971. DM 22.–

Vol. 196: H-spaces – Neuchâtel (Suisse)- Août 1970. Edited by F. Sigrist, V, 156 pages. 1971. DM 16,–

Vol. 197: Manifolds – Amsterdam 1970. Edited by N. H. Kuiper. V, 231 pages. 1971. DM 20,–

Vol. 198: M. Hervé, Analytic and Plurisubharmonic Functions in Finite and Infinite Dimensional Spaces. VI, 90 pages. 1971. DM 16.–

Vol. 199: Ch. J. Mozzochi, On the Pointwise Convergence of Fourier Series. VII, 87 pages. 1971. DM 16,–

Vol. 200: U. Neri, Singular Integrals. VII, 272 pages. 1971. DM 22,–

Vol. 201: J. H. van Lint, Coding Theory. VII, 136 pages. 1971. DM 16,–

Vol. 202: J. Benedetto, Harmonic Analysis on Totally Disconnected Sets. VIII, 261 pages. 1971. DM 22,–

Vol. 203: D. Knutson, Algebraic Spaces. VI, 261 pages. 1971. DM 22,–

Vol. 204: A. Zygmund, Intégrales Singulières. IV, 53 pages. 1971. DM 16,–

Vol. 205: Séminaire Pierre Lelong (Analyse) Année 1970. VI, 243 pages. 1971. DM 20,–

Vol. 206: Symposium on Differential Equations and Dynamical Systems. Edited by D. Chillingworth. XI, 173 pages. 1971. DM 16,–

Vol. 207: L. Bernstein, The Jacobi-Perron Algorithm – Its Theory and Application. IV, 161 pages. 1971. DM 16,–

Vol. 208: A. Grothendieck and J. P. Murre, The Tame Fundamental Group of a Formal Neighbourhood of a Divisor with Normal Crossings on a Scheme. VIII, 133 pages. 1971. DM 16,–

Vol. 209: Proceedings of Liverpool Singularities Symposium II. Edited by C. T. C. Wall. V, 280 pages. 1971. DM 22,–

Vol. 210: M. Eichler, Projective Varieties and Modular Forms. III, 118 pages. 1971. DM 16,–

Vol. 211: Théorie des Matroïdes. Edité par C. P. Bruter. III, 108 pages. 1971. DM 16,–

Vol. 212: B. Scarpellini, Proof Theory and Intuitionistic Systems. VII, 291 pages. 1971. DM 24,–

Vol. 213: H. Hogbe-Nlend, Théorie des Bornologies et Applications. V, 168 pages. 1971. DM 18,–

Vol. 214: M. Smorodinsky, Ergodic Theory, Entropy. V, 64 pages. 1971. DM 16,–

Vol. 215: P. Antonelli, D. Burghelea and P. J. Kahn, The Concordance-Homotopy Groups of Geometric Automorphism Groups. X, 140 pages. 1971. DM 16,–

Vol. 216: H. Maaß, Siegel's Modular Forms and Dirichlet Series. VII, 328 pages. 1971. DM 20,–

Vol. 217: T. J. Jech, Lectures in Set Theory with Particular Emphasis on the Method of Forcing. V, 137 pages. 1971. DM 16,–

Vol. 218: C. P. Schnorr, Zufälligkeit und Wahrscheinlichkeit. IV, 212 Seiten 1971. DM 20,–

Vol. 219: N. L. Alling and N. Greenleaf, Foundations of the Theory of Klein Surfaces. IX, 117 pages. 1971. DM 16,–

Vol. 220: W. A. Coppel, Disconjugacy. V, 148 pages. 1971. DM 16,–

Vol. 221: P. Gabriel und F. Ulmer, Lokal präsentierbare Kategorien. V, 200 Seiten. 1971. DM 18,–

Vol. 222: C. Meghea, Compactification des Espaces Harmoniques. III, 108 pages. 1971. DM 16,–

Vol. 223: U. Felgner, Models of ZF-Set Theory. VI, 173 pages. 1971. DM 16,–

Vol. 224: Revêtements Etales et Groupe Fondamental. (SGA 1). Dirigé par A. Grothendieck XXII, 447 pages. 1971. DM 30,–

Vol. 225: Théorie des Intersections et Théorème de Riemann-Roch. (SGA 6). Dirigé par P. Berthelot, A. Grothendieck et L. Illusie. XII, 700 pages. 1971. DM 40,–

Vol. 226: Seminar on Potential Theory, II. Edited by H. Bauer. IV, 170 pages. 1971. DM 18,–

Vol. 227: H. L. Montgomery, Topics in Multiplicative Number Theory. IX, 178 pages. 1971. DM 18,–

Vol. 228: Conference on Applications of Numerical Analysis. Edited by J. Ll. Morris. X, 358 pages. 1971. DM 26,–

Vol. 229: J. Väisälä, Lectures on n-Dimensional Quasiconformal Mappings. XIV, 144 pages. 1971. DM 16,–

Vol. 230: L. Waelbroeck, Topological Vector Spaces and Algebras. VII, 158 pages. 1971. DM 16,–

Vol. 231: H. Reiter, L¹-Algebras and Segal Algebras. XI, 113 pages. 1971. DM 16,–

Vol. 232: T. H. Ganelius, Tauberian Remainder Theorems. VI, 75 pages. 1971. DM 16,–

Vol. 233: C. P. Tsokos and W. J. Padgett. Random Integral Equations with Applications to Stochastic Systems. VII, 174 pages. 1971. DM 18,–

Vol. 234: A. Andreotti and W. Stoll. Analytic and Algebraic Dependence of Meromorphic Functions. III, 390 pages. 1971. DM 26,–

Vol. 235: Global Differentiable Dynamics. Edited by O. Hájek, A. J. Lohwater, and R. McCann. X, 140 pages. 1971. DM 16,–

Vol. 236: M. Barr, P. A. Grillet, and D. H. van Osdol. Exact Categories and Categories of Sheaves. VII, 239 pages. 1971, DM 20,–

Vol. 237: B. Stenström. Rings and Modules of Quotients. VII, 136 pages. 1971. DM 16,–

Vol. 238: Der kanonische Modul eines Cohen-Macaulay-Rings. Herausgegeben von Jürgen Herzog und Ernst Kunz. VI, 103 Seiten. 1971. DM 16,–

Vol. 239: L. Illusie, Complexe Cotangent et Déformations I. XV, 355 pages. 1971. DM 26,–

Vol. 240: A. Kerber, Representations of Permutation Groups I. VII, 192 pages. 1971. DM 18,–

Vol. 241: S. Kaneyuki, Homogeneous Bounded Domains and Siegel Domains. V, 89 pages. 1971. DM 16,–

Vol. 242: R. R. Coifman et G. Weiss, Analyse Harmonique Non-Commutative sur Certains Espaces. V, 160 pages. 1971. DM 16,–

Vol. 243: Japan-United States Seminar on Ordinary Differential and Functional Equations. Edited by M. Urabe. VIII, 332 pages. 1971. DM 26,–

Vol. 244: Séminaire Bourbaki – vol. 1970/71. Exposés 382-399. IV, 356 pages. 1971. DM 26,–

Vol. 245: D. E. Cohen, Groups of Cohomological Dimension One. V, 99 pages. 1972. DM 16,–